INTEGRATING SCIENCE, TECHNOLOGY, ENGINEERING, AND MATHEMATICS

How can curriculum integration of school science with the related disciplines of technology, engineering, and mathematics (STEM) enhance students' skills and their ability to link what they learn in school with the world outside the classroom?

Featuring actual case studies of teachers' attempts to integrate their curriculum, their reasons for doing so, how they did it, and their reflections on the outcomes, this book encourages science educators to consider the purposes and potential outcomes of this approach and raises important questions about the place of science in the school curriculum. It takes an honest approach to real issues that arise in curriculum integration in a range of education contexts at the elementary and middle school levels. The clear documentation and critical analysis of the contribution of science in curriculum integration—its implementation and its strengths and weaknesses—will assist teachers, science educators, and researchers to understand how this approach can work to engage students and improve their learning, as well as how it does not happen easily, and how various factors can facilitate or hinder successful integration.

Léonie Rennie is a Professor in the Science and Mathematics Education Centre at Curtin University in Western Australia.

Grady Venville is Professor of Science Education, Graduate School of Education, University of Western Australia.

John Wallace is a Professor at the Ontario Institute for Studies in Education, University of Toronto.

Teaching and Learning in Science Series
Norman G. Lederman, Series Editor

Abell/Appleton/Hanuscin: *Designing and Teaching the Elementary Science Methods Course*
Akerson (ed.): *Interdisciplinary Language Arts and Science Instruction in Elementary Classrooms: Applying Research to Practice*
Linder/Östman/Roberts/Wickman/Erickson/MacKinnon (eds): *Exploring the Landscape of Scientific Literacy*
Rosenblatt: *Rethinking the Way We Teach Science: The Interplay of Content, Pedagogy, and the Nature of Science*
Wickman: *Aesthetic Experience in Science Education: Learning and Meaning-Making as Situated Talk and Action*

Visit www.routledge.com/education for additional information on titles in the Teaching and Learning in Science Series

INTEGRATING SCIENCE, TECHNOLOGY, ENGINEERING, AND MATHEMATICS

Issues, Reflections, and Ways Forward

Edited by
*Léonie Rennie, Grady Venville,
and John Wallace*

Routledge
Taylor & Francis Group
NEW YORK AND LONDON

First published 2012
by Routledge
711 Third Avenue, New York, NY 10017

Simultaneously published in the UK
by Routledge
2 Park Square, Milton Park, Abingdon, Oxon OX14 4RN

Routledge is an imprint of the Taylor & Francis Group, an informa business

© 2012 Taylor & Francis

The right of the editors to be identified as the authors of the editorial material, and of the authors for their individual chapters, has been asserted in accordance with sections 77 and 78 of the Copyright, Designs and Patents Act 1988.

All rights reserved. No part of this book may be reprinted or reproduced or utilized in any form or by any electronic, mechanical, or other means, now known or hereafter invented, including photocopying and recording, or in any information storage or retrieval system, without permission in writing from the publishers.

Trademark notice: Product or corporate names may be trademarks or registered trademarks, and are used only for identification and explanation without intent to infringe.

Library of Congress Cataloging in Publication Data
Integrating science, technology, engineering, and mathematics : issues, reflections, and ways forward / edited by Léonie Rennie, Grady Venville, John Wallace.
 p. cm. -- (Teaching and learning in science series ; 6)
Includes bibliographical references and index.
1. Technical education. 2. Science--Study and teaching. 3. Mathematics--Study and teaching. 4. Interdisciplinary approach in education. I. Rennie, Léonie J. II. Venville, Grady Jane. III. Wallace, John (John William), 1947-
T65.I584 2012
607.1--dc23
2011047590

ISBN: 978-0-415-89756-3 (hbk)
ISBN: 978-0-415-89757-0 (pbk)
ISBN: 978-0-203-80389-9 (ebk)

Typeset in Bembo
by Taylor & Francis Books

Printed and bound in the United States of America by
Walsworth Publishing Company, Marceline, MO.

CONTENTS

Preface vii
Acknowledgements x

1 Exploring Curriculum Integration: Why Integrate? 1
  Léonie Rennie, John Wallace, and Grady Venville

2 Focus on Learning: Building Rockets and Submarines at
  Leaside High School 12
  Fiona Budgen

3 Focus on Problem-solving: Modeling an Ice Hockey Rink at
  Greenwich Public School 24
  Sheryl MacMath

4 Focus on Engineering: Bridge Building at Southern High
  School 34
  Grady Venville

5 Focus on Literacy: Linking Language and Horticulture at
  Seaview Community School 45
  Susan Joan Gribble and Léonie Rennie

6 Focus on Reinforcement: Exploring Electricity and Energy
  Use at Beachville High School 54
  Sheryl MacMath

7 Focus on Focus: Making and Marketing a Toy at Rinkview
  Public School   64
  John Wallace

8 Focus on Teacher Support: Considering Access for the Disabled
  at Gosport Community School   76
  Rachel Sheffield

9 Focus on Leadership: Constructing a Model House at Mossburn
  School   88
  Rachel Sheffield

10 Focus on Community: Learning About Tiger Snakes at Chelsea
   Elementary School   100
   Rekha B. Koul and Rosemary Sian Evans

11 Focus on Values: Investigating Water Quality in a Local Lake
   at Kentish Middle School   112
   Grady Venville

12 Reflecting on Curriculum Integration: Seeking Balance
   and Connection Through a Worldly Perspective   123
   Léonie Rennie, Grady Venville, and John Wallace

List of Contributors   143
Index   145

PREFACE

This book is about connection and relevance; it is about how we can deliver curricula that enable students to connect what they learn in their school classrooms to their world outside of school. Our focus is particularly on the school subjects of science, technology, engineering, and mathematics, the so-called STEM subjects, because we believe that these are the subjects needed for clever and creative solutions to the issues facing our rapidly changing, global world. In the real world, problems cannot be solved by experts in just one discipline, such as mathematics or chemistry; they require interdisciplinary teams to work toward solutions. For example, a real-world problem of an endangered species might be addressed by a team including zoologists, geneticists, mathematical modelers, ecologists, and geographers. In school, we believe that integrating at least some parts of the curriculum offers teachers and students opportunities to address real-world problems. Further, it enables the school to connect with its community, and thus reflect the fact that we live in a connected, global world.

Our aim in preparing this book is to encourage educators to consider the purposes and potential outcomes of curriculum integration as a way of enhancing students' skills and ability to link what they learn in school with the world outside the classroom. Curriculum integration is an educational strategy that has been in and out of popularity in cycles for many years. In the last decades of the 20th Century, it was widely promoted as a middle school strategy aimed at overcoming what seemed to be widespread disaffection of early adolescents in schools. This is still a valid strategy, but, now, its value is more likely to be argued in terms of its contribution to citizenship by offering a curriculum that is meaningfully embedded in the everyday world of students.

We regard this volume as a source book for STEM teachers and other educators who wish to be informed about different kinds of approaches to curriculum integration and the contributions they can make to the education of our students. The majority of the chapters are case stories of teachers' efforts to integrate part of their curriculum in real classrooms. These are true stories derived from a long period of research and reflection. Some teachers were more successful than others, some of their efforts were satisfying for the participants, some created situations that were tense and often stressful, but all resulted in powerful learning for the teachers involved. We think that reading about real teachers, their plans, their good intentions, the things that worked and the things that didn't, is an effective way to learn that curriculum integration is not easy, but can be very rewarding. Finding out what others have done, and thus being forewarned of the difficulties, is a fine start to understanding curriculum integration, and trying to integrate curriculum in your own classroom.

This volume is unusual for an edited book. It derives from 15 years of research by the editors, who worked on two continents with a number of colleagues, including all of the case story authors. This collaborative and coordinated approach has enabled the editors to prepare a comprehensive and cohesive volume. The authors prepared their chapters based on the elementary and middle school classrooms they researched in depth, and each case story has been chosen to highlight one, or sometimes more, important aspects of curriculum integration. All of us have taken an honest approach by discussing and illustrating both the benefits and problems associated with attempting to integrate STEM curricula. To assist readers to think about the issues that are raised in each chapter, we provide focus questions that might be used for self-study, or as a basis for graduate class discussion. We also suggest further reading of other articles or books that provide different or additional perspectives on the information presented in the chapter.

The book is organized in three sections. The introductory chapter provides a contextual overview of curriculum integration as a way to organize instruction in STEM subjects and outlines the research program that developed the case stories that follow. We point out that curriculum is integrated when it focuses on problems and issues that derive from the real world, and brings more than one subject to bear on the topic of interest, enabling students to connect their experiences inside and outside of school. We introduce six ways of integrating curricula that we found in our research program, and overview the nature of the following case stories.

The middle section of the book comprises ten case stories. The case stories presented have been carefully selected to cover a range of issues and dilemmas faced by teachers and other educators who wish to pursue an integrated curriculum, or even just a small section of integrated curriculum. Each chapter relates the story of a teacher or teachers explaining their reasons for attempting to integrate their curriculum, how they did it, and an analysis of the outcomes. We have named each chapter according to the particular focus relating to integration that is developed in the case story. We use pseudonyms for teachers and schools, but each case story is true. The ten case story chapters are arranged to build outwards, starting from the classroom and specific teaching, learning, and content issues in the early chapters through teacher professionalism and leadership issues to broader community and values issues in the later chapters.

The final chapter summarizes the important lessons learned from the ten case stories, and draws out the themes that underpin these integration efforts. Like the first chapter, this one draws extensively from the literature in order to set the findings of the case stories into a broader perspective, but also to look for ways forward. We suggest that an effective means of facilitating curriculum integration is to incorporate important local issues into the classroom curriculum and encourage students to connect these to larger, more global issues beyond the local community. This will require finding a balance between teaching to promote learning in, between, and beyond the disciplines, that is, learning that is more integrated with the world in and out of school.

Overall, these twelve chapters will assist teachers, STEM educators, and researchers to understand how curriculum integration can work to engage students and improve their learning, but also that it does not happen easily. We use the case stories to demonstrate the complexities of curriculum integration, the factors that hinder and the factors that facilitate teachers' efforts to present to their students a forward-thinking curriculum, providing them with relevant and worthwhile knowledge in today's global world. We conclude by arguing for a more holistic, Worldly Perspective on curriculum, whereby both disciplinary and integrated perspectives are employed to help solve STEM-related problems. This approach also emphasizes the connection between local issues and global concerns, enabling students' learning to go beyond cognitive, conceptual outcomes to include the social processes and real-world contexts that enable students to become effective citizens.

Léonie Rennie
Grady Venville
John Wallace

ACKNOWLEDGEMENTS

The research reported in this book was carried out with funding from three major research grants: an Australian Research Council Collaborative Grant (C59700325) with the Education Department of Western Australia, an Australian Research Council Discovery Grant (DP0451818), and a Canadian Social Science and Humanities Research Council Standard Grant (410-2006-2443). We also acknowledge the Schools, Community, and Industry Partnerships in Science (SCIps) Project that supported the Living with Tiger Snakes program at Chelsea Elementary School, and the Water Quality at Lake Wonthella program at Kentish Middle School. The Kentish program was also supported by the Australian School Innovation in Science, Technology and Mathematics Project. Both projects were funded by the Australian Government Department of Education, Science and Technology. The opinions expressed are those of the authors and should not be attributed to the funding bodies.

We greatly appreciate our colleagues who worked with us on these projects and who have contributed chapters to this book. We thank them for their insights and support. We are indebted to the many teachers and students with whom we have worked over the past 15 years, and thank them for their willingness to be involved in our explorations of integrated curricula.

1
EXPLORING CURRICULUM INTEGRATION

Why Integrate?

Léonie Rennie, John Wallace, and Grady Venville

Introduction

We live in a connected, global community. Issues involving science, technology, engineering, and mathematics (STEM) are commonplace in the media. Newsworthy events from all corners of the Earth are relayed immediately to our place of residence through a world-wide communication network. Real-time images of natural disasters, conflicts within and between countries, politicians and other leaders discussing issues relating to finance, terrorism, and human suffering, are flashed to our digital screens. Teenagers and adults alike communicate instantly with words and pictures on hand-held devices that not long ago were the province of science fiction. So much technological advancement; yet the future seems increasingly uncertain. All countries experience natural disasters of some kind; the destructive effects of earthquakes, floods, hurricanes, and wild fires are compounded by human suffering from hunger, disease, and lack of shelter. Further, no country is immune from terrorism and political unrest, or the threat created by hacking into the electronic communication networks on which we increasingly depend. We cannot predict with much accuracy the kind of world that children will occupy in the next decade.

We might hope that school education will reflect these rapid global changes and the continuing challenges relating to health and the environment. We might hope that the curriculum will foster the kind of connectedness that reflects the way the world works outside of school and assist students to develop the knowledge and ability to deal with change and challenge in sensible ways. We might expect that, as the world changes, our school systems will respond accordingly, ensuring that our youth leave school as informed citizens of the world. But we know that school systems do not adapt quickly to changing needs and, very often, curricula do not encourage such an outward-looking focus.

How can we ensure that educational curricula are designed and enacted to ready our children for the changing world outside of school? This is an old question. In 1949, Ralph Tyler wrote a classic text about curriculum entitled *Basic Principles of Curriculum and Instruction*. In it he asked four basic questions about curriculum: "What educational purposes should the school seek to attain?" "How can learning experiences be selected which are likely to be useful in attaining these objectives?" "How can learning experiences be organized for effective instruction?" and "How can the effectiveness of learning experiences be evaluated?" These remain the four questions to which educators still struggle to find answers, as illustrated by the variety of curricula evident in various countries and the diversity of views about their suitability.

The Structure of School Curricula

Now, well into the 21st century, the structure of school curricula is dominated by disciplines such as physics, mathematics, history, and literature (Scott, 2008). This pattern has existed for many decades, in the sense that disciplines, such as chemistry, physics, and biology, are readily recognized as subjects in school timetables from the past and present. Support for a disciplinary approach to curriculum flows from the belief that disciplines, like those of science and mathematics, provide specialized knowledge that enables rigorous explanation of clearly identified aspects of the world. It is argued that disciplines create a sense of order about the complex world and provide students with the understandings and skills they need to solve focused discipline-based problems (Gardner, 2004). However, most of the world's problems do not involve only one discipline; they involve many. Put another way, most school curricula are based on separate disciplines, but the world's problems are multidisciplinary and require input from many disciplines to solve them. There is another concern about the dominance of disciplines in school curricula. Rogers (1997) argued that the translation of "knowledge" from the discipline to the school subject is problematic, because the knowledge-building processes of the discipline are often lost, leaving content—the concepts and facts—disconnected from the modes of inquiry that established them. The result is a sterile, dehumanized science content that has little appeal to students and is often perceived by them to be irrelevant (Aikenhead, 2006). Further, this kind of curriculum is far removed from the practice of science in the "real world" (Hodson, 1998).

Dividing knowledge into subjects for teaching and learning in schools is a practical way to organize school timetables and deliver the curriculum (Hatch, 1998), but it is not necessarily an effective way of teaching students about the world outside of school. There are many progressive school programs that do not fit with the usual disciplinary structure of curricula and deliver programs that are "integrated," or interdisciplinary. Examples of such curricula include contextualized instruction (e.g., Rivet & Krajcik, 2008), futures studies (e.g., Lloyd & Wallace, 2004), holistic education (e.g., Miller, 2007), place-based education (e.g.,

Gruenewald & Smith, 2008), and science, technology, society, and environment (e.g., Pedretti & Nazir, 2011).

What these curricula have in common is a breaking down of the boundaries between traditional school subjects, bringing more than one subject to bear on the topic of interest. These curricula are described as integrated, deriving from the Latin word *integratus*, meaning to combine, to make whole, or bring together. Arguments supporting an integrated approach flow from the notion that curriculum should be about the problems, issues, and concerns posed by life itself. In the context of middle schools, proponents of integration argue that adolescents' learning should be about life experiences in familiar, local contexts, as well as issues in the larger, global world. Because it is based on issues of personal and social concern, such a curriculum is argued to be more relevant and motivating for students (Beane, 1995). This more holistic view enables the connections between local concerns and global problems to become a central part of the curriculum. Using disciplinary knowledge with narrow reasoning processes is inconsistent with this way of understanding knowledge (O'Loughlin, 1994); instead the disciplines become helpful only when they impinge on these issues and problems. Thus we see that having an integrated curriculum does not "do away" with the disciplines; rather they are called upon when needed to provide their specialized knowledge in context.

With these points of view in mind, and in the context of Tyler's (1949) first question, "What educational purposes should the school seek to attain?" we might ask two further questions: Do students need strong disciplinary knowledge to cope in today's changing world, or do they need cross-disciplinary, integrated knowledge? Should students learn about local issues, or focus on issues of global significance? We think the answer to these questions is "all of these things." The curriculum challenge is to strike an appropriate balance between disciplinary and integrated knowledge, and build the connections between local and global issues. How might we do this?

In subsequent chapters of this book, we present case stories of innovative teachers who have tried to find their own answers to some of these questions in the context of STEM curricula. Each of the case stories is independent of the others; they come from urban, rural, and remote areas, and from countries in the southern and northern hemisphere. The programs described in these case stories are linked together by the theme of curriculum integration involving at least one, but usually more, of the STEM subjects. All of these cases were part of a research program that began about 15 years ago, when we noticed an increasing interest in integrated curriculum, particularly at the middle school level. We noticed that many teachers were varying their science curriculum in ways they described as integrated. What did they hope to achieve by this? What did their integrated curriculum look like? What did it mean for the kinds of knowledge imparted to students, compared with the usual, subject-based curriculum? How would these kinds of knowledge prepare students to cope in the world outside of school? What could an integrated curriculum contribute to education for global citizenship? In the following section we describe how we tried to answer these questions.

Exploring Integrated Curriculum

Over the subsequent 15 years we have tried to address these and other questions about the nature and outcomes of integrated curricula (Rennie, Venville, & Wallace, 2011; Venville, Wallace, Rennie, & Malone, 2002). Perhaps not surprisingly, we found that teaching and learning are complex processes, made more so by the individuality of each teacher and his or her classroom. At the risk of simplifying these complexities, we suggest four generalizations about the nature and purpose of integration in the STEM curricula we examined.

The first generalization is that integrated curricula come in many shapes and forms. In all cases, however, there was an attempt to lead students to look outward toward the real world while drawing from the strengths of the subject disciplines. The second, and related, generalization is that when teaching higher grades, particularly, teachers had to put in considerable effort, in terms of both energy and time, to overcome the subject-centered structure of the school curriculum.

Our third generalization is that integration typically occurred within a context relating to the outside world and students' personal experience. If we think of learning as making meaning, then this context is "the frame of reference that provides meaning" (Clark, 1997, p. 71). Making meaning requires the learner to make connections between things, by making patterns, organizing experiences, and creating "meaningful wholes." As Clark (1997, p. 70) suggested, "an integrated curriculum begins with an assumption of the 'connectedness of things,'" and further, that "an integrated curriculum is learner-centered." This leads us to the fourth generalization: The degree of integration is related to the degree of learner-centeredness in the curriculum. We found that some curricula were primarily teacher-centered, but there were still important parts over which learners could exercise independence and control. Where this happened, learners were able to find ways of connecting local experience to a bigger, more global picture.

One issue related to curriculum integration is that of terminology. There is a range of terms used to describe integrated practice. For example, the term "integrated curriculum" is sometimes used interchangeably with "interdisciplinary curriculum." Other words often used to describe integrated curricula include multidisciplinary and transdisciplinary. Clearly, all of these words refer to how disciplinary knowledge and skills are dealt with in an integrated curriculum. In this volume we use these terms in a way that is consistent with their general use:

Unidisciplinary describes a curriculum that draws from a single discipline.

Multidisciplinary programs draw from more than one discipline, each contributing to the topic of interest, but these disciplines remain separate.

Interdisciplinary curricula also draw from more than one discipline, but here the boundaries between these different disciplines become blurred or broken down as they are used to tackle a particular topic or theme.

Transdisciplinary curricula are fully integrated in that they draw from several disciplines, using the knowledge most appropriate to deal with the problem or issue at hand. The disciplines merge into each other, boundaries disappear, and the curriculum tends to be structured around the needs of the learner, rather than preserving the structure of the various disciplines.

The distinctions we make above have been used by some writers to create a continuum of curriculum integration, a kind of progression from less to more integrated. Drake (1993), for example, structured her three-framework continuum quite simply, taking the terms multidisciplinary, interdisciplinary, and transdisciplinary as frameworks. Earlier, Jacobs (1989) used a five-part continuum, beginning with "parallel discipline designs," where teachers of different subjects arranged to teach similar material in parallel so that there was reinforcement of the subject matter, but the subjects remained separate. Second was "complementary discipline units," similar to the descriptions of multidisciplinary courses, followed by "interdisciplinary units/courses." Both of these approaches maintained the structure of the disciplines in some measure. Jacobs's "integrated day" described a theme-based program focused on students' interests. The fifth kind of integration was described by Jacobs as a "complete program," in which students determined their own curriculum according to their needs and interests.

It is easy to find continua like these and some even more complex (see, for example, Fogarty, 2002). What these curricula have in common is the naturally bounded school subjects as the starting point for curriculum organization, graduating into a curriculum in which the subject structures are broken down and the curriculum is organized around "real world" problems, issues, or projects of relevance to the learners.

These continua may be helpful in understanding how curriculum integration may vary, but we find them problematic. We find that the notion of continua implicitly suggests that moving along a continuum represents progress from a less desirable to a more desirable position; that more integration is better, for example. This approach is at odds with our research findings. Certainly there were different degrees to which subject boundaries were broken down and the curricula were more or less student-centered, but we found no inherent quality of "betterness." Instead, we believe that the effectiveness of each curriculum must be judged according to its purpose. We agree with Hargreaves, Earl, and Ryan (1996), who pointed out that a continuum does not capture the complexity of integration. They recommended taking a pragmatic position that acknowledges and incorporates many different forms of integration. In our view, this position better reflects the broad spectrum of implemented curricula. To this end we have suggested an inclusive description of integrated curriculum as one that enables students to look toward multiple dimensions that reflect the realities of their experiences outside and inside school (Venville, Rennie, & Wallace, 2012).

Approaches to Integrating the Curriculum

Our research revealed a diversity of teachers' efforts to integrate curricula, with many different approaches to integration. In describing the different approaches, we chose to use terms that teachers themselves used. Terms such as thematic and project-based provide information about how the curricula were structured, how they were implemented in the schools, and the relationship between subjects. Here, we describe six approaches to integration and illustrate them with examples, in the form of case stories, from our research program. Pseudonyms are used for all schools and teachers. The approaches are broadly sequenced in terms of whether the subjects are taught separately or together, but there is considerable overlap among them; none should be regarded as a "pure" form of curriculum integration. The order is one of convenience; no hierarchy is intended. The six approaches are synchronized, thematic, project-based, cross-curricular, school-specialized, and community-focused.

Synchronized Approach

Synchronized approaches are based on the identification of specific skills, knowledge, or understandings that are part of the content of more than one subject. These parts of the curriculum remain separately taught, although often at similar times, so that the presentation of content is in parallel, or synchronized, between the different subjects. Teachers from different subject areas identify points where connections may be made between pre-existing topics, they explicitly draw the links in their classes and teach them in a similar manner, sometimes using common tasks or assignments. This is an example of a multidisciplinary approach; the subjects remain separate, but the common links are taught in parallel. For example, at Beachville Public School, the case story in Chapter 6, the science and geography teachers were dealing with units in electricity and energy use at about the same time with the same class. With thoughtful planning, they identified specific links between the content in each subject and then ensured that these specific links were taught synchronously, with the intention that two perspectives of similar content would assist and reinforce students' understanding.

Thematic Approach

Thematic approaches organize the curriculum around a particular local or global topic that enables the content of different subject areas to be linked together. The integrated unit might run for a period as short as a single day of themed activities, so as to celebrate an event or festival for example, or for a term, where several subjects are taught in complementary ways that relate to the theme. Sometimes classes are brought together for a culminating thematic event, such as an excursion or a special day of activities when work is displayed. Often, the subjects are taught separately in different classrooms, with teachers and students expected to make the connections back to the theme, so this can be described as a multidisciplinary approach. At Gosport Community School, our Chapter 8 case story, two teachers used the theme of community access for disabled people to bring together aspects of science, social studies, and health with their classes of grade 8 students. Although teachers taught their subjects separately, the content was planned to highlight access issues for disabled people.

Project-based Approach

Project-based approaches to integration are built around a planned task, often based on technology or engineering, requiring knowledge and skills from more than one subject area for its completion. The subject content is applied as and when it is needed to complete the project, and because the focus is on the project, the boundaries between the separate subjects become blurred and less visible. The project-based approach differs from the thematic approach, where the subjects usually remain separate from each other, although the knowledge and skills from each are used in a complementary way. Whereas the thematic approach might be described as a multidisciplinary approach, the project-based approach is inter-disciplinary. The subjects have more interconnections than in a theme, and the connecting links are made clearly to the students. In Chapter 4 we report a case story at Southern High School, where grade 9 students were challenged to build a model bridge in their technology class. The students drew on knowledge from science and mathematics to build their bridge. Because they also needed to complete the bridge on budget and on schedule, the project also involved principles of engineering, aesthetics in design, and aspects of economics.

Cross-curricular Approach

A cross-curricular approach to integration is more pervasive than the approaches mentioned so far. This approach is usually based around overarching skills or competencies, such as literacy or numeracy, or on values or social skills, such as environmental responsibility or cooperation and teamwork, that cut across a number of subject areas. Because these skills are the focus of more than one subject at the same time, integration can occur across these subjects. For example, incorporating communication technology might be used as an integrating subject across others, such as design technology, science, mathematics, and art. Skills and competencies related to keyboarding and the use of various software blur the boundaries between subjects, especially the boundaries with technology. An excellent example of blurring of boundaries was achieved at Seaview Community School, our Chapter 5 case story, that used literacy in a cross-curricular approach to the whole-school curriculum as a means of assisting students who did not have English as their first language.

School-specialized Approach

Some schools specialize by having a long-term focus on a particular subject area, such as engineering, or perhaps a sport, such as basketball. Such a curriculum focus may offer tuition in its subject specialty, but a school-specialized approach to integration means that this focus is embedded in the whole school curriculum, and staff in the school will tailor a range of other subjects to have explicit links to the chosen specialization. For example, a school that specializes in engineering may have courses or electives in high-level mathematics and science that support its courses in engineering, and perhaps have enrichment or advanced-placement programs related to engineering for students. The school may also have open days or other ways of advertising its special focus. There will be teachers and school infrastructure that support the specialization because it is a long-term commitment. One of the high schools in our research program, located near the coast, offered a school specialization in marine studies. At this school, aspects of marine science were integrated into other school subjects, such as science (marine biology), social studies (ocean weather and currents in geography as well as social and historical aspects of the port and fishing industry), mathematics (navigation), and technology (boat-design and engineering).

Community-focused Approach

Some approaches to integration are based on issues that have significance in the school's local community. When the curriculum involves an in-depth investigation of a local problem or issue, we describe it as a community-focused approach. Here, school subjects serve to help students understand the issue, and to find and review potential solutions. There may be a culminating event or other tangible outcome, in a similar way to the project-based approaches; however, a community-focused approach often results in some action by the students that has implications for the community. Students may prepare an information brochure, or take some initiative related to the environment, giving them an outward-looking focus that goes beyond the constraints of school subjects. Two of our case story chapters, featuring Chelsea Elementary School (Chapter 10) and Kentish Middle School (Chapter 11), describe community-focused programs involving studies of nearby lakes. The schools adopted different approaches, with Chelsea students learning about lake ecology and finding ways to educate the community about venomous snakes that lived around their wetland, and Kentish students investigating factors that concerned the ecological health of their local lake. Community-focused approaches can be described as interdisciplinary, but they also have the potential to be transdisciplinary. Because they originate from real-life community problems there is a compelling context for learning about, and making meaning from, the issue itself, rather than just drawing from one subject or another.

An Invitation

For this volume, we have selected ten case stories from our larger research program to demonstrate a variety of reasons for teachers choosing to integrate, and a range of ways in which they did so. We describe what teachers did, and discuss both the benefits and problems they experienced. Some case stories focus on one of the important reasons for, or means of, choosing to integrate the curriculum. Others focus on the factors that facilitate or hinder the success of the integrated program. Our aim in this volume is to illustrate integrated curriculum in the STEM subjects at the elementary and middle school levels so that you can see what others have done, try for yourself, and be prepared for the challenges ahead in your curricular adventure. We invite you to read the book, and learn from our experience and the experience of the teachers and students in our case stories.

Focus Questions

1. Think back on your own schooling in the STEM subjects. What kinds of lesson content did you enjoy the most? Why was this? Can you identify aspects of these courses that were integrated? What were the most and least enjoyable aspects?
2. Have you previously tried to integrate some curriculum content in your own classroom? If so, what were the most rewarding outcomes of your efforts? What were the greatest challenges you needed to overcome?
3. Linking in-school STEM curriculum content to local issues in the community outside of school invariably involves an integrated approach. What are two important and topical issues in your local community at the moment? What opportunities can you think of to bring any (or all) of the STEM curricula to bear on these issues?

Suggestions for Further Reading

Clark, E. T. Jr. (1997). *Designing and implementing an integrated curriculum*. Brandon, VT: Holistic Education Press.

This book examines many of the myths and beliefs that are advanced as arguments against integrated curricula and shows how they do not hold up in a fully integrated curriculum that is learner-centered. Although published some years ago, it remains very practical, showing how a middle school curriculum can be designed around concepts and worthwhile questions to create integrated, relevant, and provocative curricula. The book also emphasizes the importance of staff development in the implementation of new curricula.

Beale, C., Grable, A., & Robertson, A. (2001). *Curriculum integration: Middle school educators meeting the needs of young adolescents*. Raleigh, NC: North Carolina State University. Accessed via www.ncsu.edu/chass/extension/ci/index.html.

This website provides an excellent and practical overview of curriculum integration. It covers the history and philosophical underpinnings of integration, and also provides a case study of one teacher's journey with integration and looks at issues such as evaluation and assessment.

References

Aikenhead, G. (2006). *Science education for everyday life. Evidence-based practice*. New York: Teachers College Press.
Beane, J. A. (1995). 'Curriculum integration and the disciplines of knowledge'. *Phi Delta Kappan*, 76(8), 616–622.
Clark, E. T. Jr. (1997). *Designing and implementing an integrated curriculum*. Brandon, VT: Holistic Education Press.
Drake, S. M. (1993). *Planning integrated curriculum: The call to adventure*. Alexandria, VA: Association for Supervision and Curriculum Development.
Fogarty, R. (2002). *How to integrate the curricula* (2nd ed.). Thousand Oaks, CA: Corwin Press.
Gardner, H. (2004). 'Discipline, understanding, and community'. *Journal of Curriculum Studies*, 36(2), 233–236.
Gruenewald, D. A. & Smith, G. A. (2008). 'Introduction: Making room for the local'. In D. A. Gruenewald & G. A. Smith (Eds.), *Place-based education in the global age: Local diversity* (pp. xiii–xxiii). New York: Lawrence Erlbaum Associates.
Hargreaves, A., Earl, L., & Ryan, J. (1996). *Schooling for change: Reinventing education for early adolescents*. London: Falmer.
Hatch, T. (1998). 'The differences in theory that matter in the practice of school improvement'. *American Educational Research Journal*, 35(1), 3–31.
Hodson, D. (1998). 'Science fiction: The continuing misrepresentation of science in the school curriculum'. *Curriculum Studies*, 6(2), 191–216.
Jacobs, H. (Ed.). (1989). *Interdisciplinary curriculum: Design and implementation*. Alexandria, VA: Association for Supervision and Curriculum Development.
Lloyd, D. & Wallace, J. (2004). 'Imagining the future of science education: The case for making futures studies explicit in student learning'. *Studies in Science Education*, 39, 139–177.
Miller, J. (2007). *The holistic curriculum* (2nd ed.). Toronto, ON: University of Toronto Press.
O'Loughlin, M. (1994). 'Being and knowing: Self and knowledge in early adolescence'. *Curriculum Perspectives*, 14(3), 44–46.
Pedretti, E. & Nazir, J. (2011). 'Currents in STSE education: Mapping a complex field, 40 years on'. *Science Education*, 95(4), 601–626.
Rennie, L. J., Venville, G., & Wallace, J. (2011). 'Learning science in an integrated classroom: Finding balance through theoretical triangulation'. *Journal of Curriculum Studies*, 43(2), 139–162.
Rivet, A. E. & Krajcik, J. S. (2008). 'Contextualizing instruction: Leveraging students' prior knowledge and experiences to foster understanding of middle school science'. *Journal of Research in Science Teaching*, 45(1), 79–100.
Rogers, B. (1997). 'Informing the shape of the curriculum: New views of knowledge and its representation in schooling'. *Journal of Curriculum Studies*, 29(6), 683–710.

Scott, D. (2008). *Critical essays on major curriculum theorists*. London: Routledge.
Tyler, R. W. (1949). *Basic principles of curriculum and instruction*. Chicago: The University of Chicago Press.
Venville, G., Rennie, L., & Wallace, J. (2012). 'Curriculum integration: Challenging the assumption of school science as powerful knowledge'. In B. Fraser, K. Tobin, & C. McRobbie (Eds.), *Second international handbook of science education* (Vol. 2, pp. 737–749). Dordrecht, The Netherlands: Springer.
Venville, G., Wallace, J., Rennie, L., & Malone, J. (2002). 'Curriculum integration: Eroding the high ground of science as a school subject?' *Studies in Science Education*, 37, 43–84.

2

FOCUS ON LEARNING

Building Rockets and Submarines at Leaside High School

Fiona Budgen

Introduction

When considering the implementation of integrated curricula in schools, it is important to examine how learning progresses in such settings. What is the influence of integrated learning on student motivation? Does integration increase the likelihood of learning opportunities becoming learning outcomes? Can teachers create a classroom environment that promotes learning within and across disciplines? This chapter describes a year-long study of engineering projects and the quest to assess their impact on a group of 19 mixed-ability grade 8 students at Leaside High School, a medium-sized school in an urban area. Each project integrated the curriculum areas of design technology with mathematics and science, and was under the direction of the same two teachers. Ms Davis was responsible for mathematics and science while Mr Batani taught design technology. As the researcher, I joined the class as a participant-observer and was actively involved with the teachers and students for the duration of the study. Rather than passively monitoring students' learning in each project, I was part of a collaborative process to find effective assessment strategies and to strive for successful outcomes. The following section provides a brief description of the four integrated design projects which set the context for the outcomes and conclusions that emerged from the study.

The Integrated Projects

Project One: Bottle Rockets

In the first project students were asked to build rockets using large plastic bottles that were propelled by using compressed air. The challenge was to build the rocket that would fly the greatest distance, and students were able to add attachments, such as fins, to assist flight. Teachers had deliberately chosen a simple task with very little design input required from the students, who were new to the high school setting, had no prior experience of working in a science laboratory or a wood- or metal-working room, and would need time to get to know their new classmates and teachers. Leaside High School had well-equipped technology rooms so students had easy access to an appropriate space for construction projects.

Students were able to make their own rockets, but they also had to form small collaborative groups of two or three because they needed to help each other during certain phases of the building process and during the launching and testing of the rockets. The technology teacher, Mr Batani, built an adjustable launch pad for the rockets so that different launching angles could be tested. To launch the rockets the bottles had to be about a third full of water. Each student held his or her rocket on the launch pad while a partner injected the bottle with compressed air through a hole in the bottle's screw cap. With sufficient compressed air inside, its expansion launched the rocket.

A simple, three-step process of make → evaluate → change was followed with none of the rockets flying straight or very far on the first attempt. Several spiraled overhead and sprayed the onlookers with water before crashing to the ground behind the launch pad. It was a case of back to the drawing board to evaluate the events and to consider what changes were needed in order to make the rockets fly straighter and further. Once the rockets did start to fly well, the students were able to measure the distance travelled and to test the effectiveness of the different launch angles. The more competent groups went on to learn how to use an inclinometer and to calculate the vertical height reached by their rockets.

Project Two: Rocket Cars

The second project involved students in designing and building model rocket cars, again using bottles powered by compressed air. The goal was to produce a vehicle capable of travelling the fastest along a 35 m horizontal guide-track. The project was more complex than the bottle rockets activity because this time a design was not provided, so students were expected to follow a four-step process of: design → make → evaluate → change. Much of the knowledge required for the first project, particularly about aerodynamics, needed to be applied to this second project.

The testing of the first prototypes revealed many design problems. Most of the vehicles travelled only a short distance before stopping or leaving the track. By the beginning of the third day of trials the students were making little headway, so Mr Batani began building a vehicle of his own that he systematically tested and changed, talking aloud as he did so. This approach produced the desired impact on the students, who stopped making random changes to their vehicles and began to work methodically. By the end of the lesson, all of the vehicles had improved and several were consistently reaching the end of the track.

Project Three: Pull-Along Toys

The third project involved designing and building wooden pull-along toys that were required to have at least two moving parts. This project differed from the previous two in that the students had much greater freedom in the design of their finished products. The design process was more complex, with students having to incorporate two moving parts using cams or eccentrics as mechanisms to create movement.

Many of the students chose to work individually because they wanted to keep their final product. Once the students had decided on a toy to build, the next stage was to create a full-size, two-dimensional mock-up using stiff paper and pin board. The mock-up had to be accurately measured and drawn because it would provide the template for the shapes to be cut from wood. Students attached each paper part in place on the pin board with thumb tacks, so that the parts could show the mechanisms for the moving parts of the toy. This allowed students the opportunity to test their ideas and explore some of the problems that they might encounter with their designs.

Design problems that were not addressed at this stage became major hurdles during the building phase; nevertheless, nearly everyone managed to complete the project. Finished toys included an Egyptian archer in a chariot, a duck, a mouse, a truck, a dolphin, and an army tank.

Project Four: Submarines

This project took place during the fourth school term. Students built and tested model submarines that had been devised and designed by Mr Batani. Unlike the previous two projects, no design input was required from the students; instead the focus was on building and performance.

The submarine was made from a plastic drink bottle with a lead keel attached. The submarine was designed to allow water to enter through an intake valve at the underside and for air to escape through a valve at the top, which was held open by a wooden float. As the water level rose in the main chamber, the submarine sank. The float holding the top valve open rose with the water, allowing the valve to close. On the top side of the submarine was a canister for dry ice. When water entered the canister, the dry ice vaporized more quickly, filling a main chamber with carbon dioxide gas and forcing the water out through the copper pipe at the back, thus propelling the submarine forward. As the submarine's buoyancy increased with the displacement of the water, the submarine rose to the surface for the cycle to repeat.

The students were not given a choice about whether they could work in groups or as individuals in this project. However, in order to build all the components in the time available they had to work as teams with each member taking responsibility for different parts so that they were ready for the scheduled testing at Leaside's swimming pool. Before they began to build, each group devoted time to preparing a production plan and deciding who would be responsible for manufacturing each of the submarine's components. The teacher prepared instruction sheets for each component—the keel, the conning tower (the canister for dry ice), the valves, and the fore and aft ballast tank—with the intention that students could work with some degree of independence.

The first tests were done without dry ice and were conducted to ensure that the submarine would sink nose-first and sit level when it settled on the bottom of the pool. Once the students were sure that their submarines were correctly trimmed, they went on to test the performance using dry ice. Most groups had to solve a few problems at the outset, but once the submarines were operating successfully, they porpoised through the water, diving and resurfacing about four or five times before having to be refueled.

Outcomes from the Integrated Projects

This chapter began with three questions about how learning progresses in an integrated setting. These questions concerned students' motivation, learning, and transfer of knowledge. These factors frame discussion of the outcomes of the four projects.

Students' Motivation

Students' levels of motivation fluctuated conspicuously and quite unpredictably at times. These fluctuations existed between projects and within projects. The observation and interview data from the first integrated project, the bottle rocket, exposed some evidence of motivation, but the reasons for it were not clear. While many of the students enjoyed the project, their reasons for enjoying it were varied. Only one of the nine students interviewed gave an answer that directly linked the integrated nature of the project to her enjoyment of it. Other motivating factors offered by the students could have occurred whether the project was integrated or not.

Evidence of motivation turned out to be equally hard to find in the second project, the rocket car. Although students were keen to get started on building their cars, they quickly lost enthusiasm when they began to encounter difficulties. A pattern appeared to emerge from the first two projects. A number of students lost interest in the first project when they could not get their rockets to fly successfully. Even though the instructions had been provided for building the rockets, these students did not appear to have the skills, patience, or persistence to apply themselves to the task of finding out where they had gone wrong and making changes in order to build a successful rocket. Instead, they tried to divert attention from their poor performance with irrelevant tactics such as adding detergent to the water in the bottle. The same thing happened with the rocket car. Enthusiasm for the project quickly evaporated for a number of students who found that the task was not as straightforward as they first thought and, once again, those who lacked the skills, patience, and persistence to problem-solve tried to disrupt the activities of others.

This pattern continued during the third project when some previously successful students found themselves faced with more challenging problems. Their motivation to tackle the challenges of the pull-along toy project quickly waned when the challenges appeared too great. This outcome points to the interconnectedness of motivation and skills, and supports an emerging perception that the signs of motivation could not be directly attributed to the integrated nature of the projects, but rather to the level at which the projects matched the students' existing skills and knowledge, and their ability to find possible solutions.

It was clear that many of the students were not used to problem-solving, and had trouble staying on task in the independent learning environment provided by the integrated projects. In order for independent learning to take place, students must feel a certain level of intrinsic motivation, that is, there must be a desire to learn, which, in turn, can be fostered by providing stimulating tasks. The teachers, Mr Batani and Ms Davis, had put considerable effort into providing a series of tasks that they hoped would stimulate intrinsic motivation and there was evidence that some students were motivated. However, a number of students did not respond to the projects in a way that indicated any level of intrinsic motivation, or readiness to take responsibility for their own learning. This was exposed when students, who generally worked quite compliantly during the more closely controlled parts of the first two projects, became foolish and disruptive during the testing and refining stages when they had greater freedom of movement and thought that they were less accountable for their actions.

The more promising indications of motivation came gradually and were revealed by changes in some students' attitudes. There were many examples of change, but probably the most striking example was from Luke who, at the beginning of the year, was one of the most troublesome characters in the class. He was by far the biggest student in the class and had an overbearing manner with others. He showed little interest in the early tasks and was the subject of complaints from other students in their reflection sheets. I interviewed Luke outside the classroom toward the end of the pull-along toy project. His responses came as quite a surprise and did not reflect the persona he had hitherto displayed in class. He revealed himself to be intelligent, articulate, and thoughtful, with a realistic understanding of the problems he and his partner were facing in their project. There was a noticeable change in Luke's behavior after the interview; he worked diligently with his group on the submarine project and showed little sign of the domineering character that had distracted and disrupted others during the previous projects. The change in Luke's behavior was evident from the moment he returned to the class and remained apparent until the end of the study. Although this was not the intention of the interview, it seemed that the process of reflecting on his activities helped Luke to recognize that the learning opportunities presented by the projects were his to take if he applied himself to the tasks.

The change in students' attitudes had certainly not happened overnight. It had taken a great deal of patience and perseverance from the teachers to persist with a program that they believed would have long-term benefits for the students in setting them on the path to becoming independent, self-motivated learners. The change in students' attitudes came late in the year and it was impossible at that stage to assess whether these changes would be lasting. However, the indications that the improvement in attitudes could partly be attributed to an increased level of intrinsic motivation was a promising sign that the integrated projects may have opened the door to a fundamental shift in the way that many of the students approached their work. Some had begun the year showing little interest in the tasks, a reluctance to work independently, and a lack of self-discipline. The manner in which they worked to complete the final project suggested that a more mature approach to learning was emerging.

Assessing Learning Outcomes from the Integrated Projects

One of the greatest challenges for teachers was to identify and assess the learning outcomes from each project. The intended learning from mathematics, science, and design technology could be identified from the curriculum, but the teachers hoped that the integrated projects would take the students beyond the curriculum. They hoped that the projects would help the students to develop skills in cooperative learning and problem-solving, and to become less dependent on the teacher as the source of all knowledge. Teachers began a journey to find effective ways to measure the learning that took place. At times it was a frustrating journey when we saw the paucity of useful information that some forms of assessment provided and when we reached dead ends with other strategies. This section describes the journey and the paradigm shift that evolved for the teachers in their choice of assessment strategies.

The assessment of the first project, the bottle rocket, was based around the work that students completed in their workbooks and observations about students' ability to work cooperatively and to follow basic workshop procedure. However, the workbooks did not provide a clear picture of each student's learning during the project because the effort put into completing the tasks was spasmodic, much of the work was done in collaboration with others, and there was no way of knowing how much of the learning was a direct result of the project, or a result of knowledge that students had brought with them. A number of the interview questions identified key concepts and skills that were intrinsic to the project. Other interview questions were more open to allow students to discuss learning that had been significant to them. The answers to these questions revealed that some of the hoped-for outcomes had occurred but were not commonplace. These assessment results exposed a need to refine the assessment strategies in order to compile a clearer picture of the learning that took place during each project.

A design portfolio was introduced during the second project to document the development of the designs for the rocket cars. Students were to detail changes they had made, explain why these changes were necessary, and record the results. It was hoped that these portfolios would provide a written account of the formative learning process for each student as he or she worked through the challenge. However, as with the bottle rocket project, the students were reluctant to keep their records up-to-date and, even though time had been set aside to work on the portfolios, they did not provide as much evidence of learning as the teachers had hoped.

At the end of the project students completed a reflection activity. The reflection sheets allowed students to focus on the concepts and skills that were significant to them and gave some insight into the nature of the learning that took place. Many students commented about the need to do more planning and to think more carefully about the design. Comments about the need to keep the vehicle light, about the shape of the vehicle, and about problems with the steering system indicated that thought had been given to the aerodynamics of the vehicle, the materials used, and the problem of friction between the steering system and the guide track. However, it became apparent that, although the students were in the care of two committed and conscientious teachers, their learning lacked specific focus and direction. Ms Davis, Mr Batani, and I sat down together before the third and fourth projects to map out the precise and measurable learning outcomes to be achieved through each project. This marked a paradigm shift for both teachers. Rather than making statements about what the students would do they began to identify what they expected the students to learn.

Another form of assessment was required that would enable a comparison of learning before and after the project so it was decided to try concept mapping. This skill was new to the students and time had to be spent explaining how to create a concept map before using the technique as an assessment tool in the third project of designing and building a pull-along toy. However, promising indications that concept maps could be an effective assessment tool were not fulfilled. In fact, the only thing that was reliably indicated was which students were becoming competent in the techniques of creating a concept map. A great deal more practice would have been needed before all of the students were sufficiently skilled in creating concept maps that validly reflected their knowledge. With only one more project to be completed, it seemed unlikely that the students' skills in creating concept maps would improve sufficiently to the point of using the maps as an assessment tool for the fourth project.

By the fourth project, there was a growing sense of the inadequacy of the assessment strategies that had already been tried. These strategies had failed to effectively identify the learning that could be attributed to each project, they had been difficult to manage, they had been difficult for the students to complete, and they had often belatedly revealed students' miscomprehensions or lack of comprehension. We found ourselves looking for an alternative assessment strategy. An assessment tool was required that would allow monitoring of students' learning as it took place. The solution was found in the use of learning journals. It was vital that students had a sound understanding of how the submarine operated before they went to the swimming pool to test their vessels and the learning journals played a crucial role in allowing teachers to check students' understanding, correct misconceptions and assist those who had not grasped particular concepts. The learning journals allowed teachers to monitor learning as it took place and to sustain understanding as the project progressed.

The use of learning journals may also have facilitated learning in another way: Giving the students the opportunity to express their understandings required them to organize their thoughts, and to use the mathematical, scientific, and technical language they had encountered during the project. Vygotsky (1986) supported the notion that "speech is an expression of that process of becoming aware" (p. 30) and although the learning journals did not initiate a great deal of speech, they provided the opportunity to use language in a similar way by creating a written dialogue between student and teacher. Other authors, such as Ben-Yehuda, Lavay, Linchevski, and Sfard (2005), have expanded on this "communicational approach to cognition" (p. 176) and would define the face-to-face interviews as synchronous discourse and the learning journals as asynchronous discourse.

As related earlier in Luke's story, the action of interviewing students and asking them to reflect on their learning may have been a catalyst for students to become more aware of their own thinking and learning processes. Encouraging students to develop skills in metacognition allows them to become independent monitors of their own progress and success (Schoenfeld, 1987).

An interesting twist in the close relationship between motivation and learning was exposed during the final project. The students began the project with very little motivation and some were openly resistant to the task that was being set for them, but by the end of the project they were engaged in their work, appeared to be motivated, and were willing to persevere during the testing stage to make their submarines operate successfully. The students also began the project by being unwilling to work with others, but by the end of the project were cooperating well with each other. As well as revealing that the students were capable of cooperative learning and some independent problem-solving, this project produced the most successful learning outcomes, yet it had a very inauspicious start. Why did the students' attitudes change during this project? What was the key to its success?

Motivation was certainly not the answer, given the students' initial responses to the project. Instead the answer may reside in the way that learning was structured in this project. There was no design phase in the submarine project; students did not have to make choices about what to build or how to build it; nor did they have to investigate how to make it work successfully. A blueprint of the design was provided for students to follow and the principles that enabled the submarine to function were explicitly taught. The students knew exactly what they had to do to build and operate their submarines. They had been provided with sufficient information to solve problems if the submarines did not operate successfully. The teachers were clear about the knowledge required to successfully complete the project, and the use of learning journals reinforced to students the learning that was important.

Transfer and Application of Knowledge

One of the key arguments for integrated curriculum is that it can promote learning that is rich with connection-making (Perkins, 1991). So, what evidence emerged from the integrated projects to indicate transfer and application of knowledge within and across learning areas?

The first project revealed only emerging signs of the flexible use of knowledge. One manifestation of this was from the students who commented that the most important thing they had learned was the need for care and accuracy in the building process; however, there were compelling signs that many students failed to transfer and apply the skills and knowledge they had used in the first project to the next project. Important learning was also not transferred across subject areas, for example, the consistent failure to apply knowledge about testing variables, the failure to recognize the implications of the track measurements for the vehicle's design, or to give consideration to the effects of friction, all indicated that connections had not been made to the content being taught in science and mathematics.

One comment that appeared regularly in the rocket racer reflection sheets was the need to take more time in designing and planning the vehicles, and this turned out to be an important feature of the third project, the pull-along toys. Students' written reflections on the reflection sheet for the pull-along toy project and interviews conducted at the end of this project revealed that many students still felt that their planning had been inadequate.

The design of the toy had to be well thought out before the building process began and the wood cut or glued. With some assistance, all students were able to create a design that would work using a paper template on pin board, but the problems in many cases arose from an inability to visualize the two-dimensional template as a three-dimensional object. The students engaged with the skill of translating two-dimensional outlines into three-dimensional objects during the mathematics lessons associated with the project; however, it was only the more able students who successfully recreated three-dimensional models from two-dimensional sketches. Since many of the students did not master this skill in mathematics, it was not surprising that they were unable to transfer the skill to their toy-making project. What was surprising, however, was that none of students who were asked about the mathematics they had used when making their toys mentioned these spatial concepts and none of the students identified this aspect of the project as being difficult unless they were specifically asked for comment.

In the final project little was left to chance. The students needed to have a thorough understanding of the operation of the submarine in order to make the vessel work successfully. It was clear from the previous projects that it could not be assumed that students would recognize the significance of learning in mathematics and science, or that they would apply that learning to the project. The essential understandings that applied to the operation of the submarine were revised, checked, and rehearsed through the learning journals until the teachers felt confident that all students had a sound understanding of the process. The students could not have achieved the success they did with their submarines had learning not been transferred and applied across subject areas, but the skills to apply their knowledge also had to be explicitly taught in order to ensure the success of the project.

Conclusions

The events during the four projects provided a complex and multi-layered text, but one thing became clear: the notion that integration could, by its very nature, motivate students, improve learning outcomes, and facilitate connections between learning areas was an inadequate and simplistic notion.

The synergy between motivation and learning was the most interesting interaction observed. While motivation is often thought to be a precursor to meaningful learning, the projects showed that meaningful learning could be a powerful factor in shaping student motivation, thus supporting the conclusion that purposeful learning enhances motivation and confirming the observation by Caskey (2002) that there is not only a need for understanding and support in middle school education, but also that academic achievement needs to be taken seriously as well. Undertaking the projects emphasized the importance of clearly defining the intended learning and conveying this information to the students. They confirmed the need to explicitly teach knowledge that is pivotal to the success of a project and showed that, when these conditions are met, learning can become both purposeful and motivational.

The second conclusion pertains to the learning that takes place in an integrated setting. It appeared from the observational data that many of the students were reluctant to problem-solve for themselves, but other evidence indicated that the students knew what they had to do; they just didn't know how to do it. This exposed the students' need to be supported in the challenge of approaching integrated learning. The projects revealed that a formative assessment tool such as the learning journal could be a potent support to learning. The learning journal also provided a crucial ingredient in supporting learning and cognitive development by making provision for the students' use of language. Doerr (1996) asserted that more time needed to be given to allow students to discuss, conjecture, and validate their ideas. The students had many opportunities to engage in kinesthetic learning activities through the projects, but it was not until these were intertwined with the use of language through the learning journals that there was a major improvement in the learning outcomes. If learning is facilitated by a combination of language and action, then the two need to be interwoven in order to improve understanding and retention of relevant concepts. The combination of action and language, and the use of integrated assessment strategies, are approaches that prepare and support the students to meet new challenges on their learning journey.

The third conclusion brings scrutiny to the effectiveness of the integrated projects in enabling students to flexibly apply their knowledge. It became evident that the network of interconnections across and within learning areas should be tightly knitted to the framework of each project. Transfer and application of knowledge could not be assumed, even when the connections appeared palpable to the adult eye. It was vital to support the students to develop the cognitive skills that allowed them to flexibly apply their knowledge. We found that the same approaches that prepared and supported the students to meet new learning challenges—that is, the combination of action and language, and the use of integrated assessment strategies—also promoted the transfer and application of learning.

Both teachers had a strong belief in the importance of bringing learning into context and identifying connections across curriculum areas. As the year progressed and it became clear that achieving the goals of an integrated curriculum was more elusive than had been anticipated, Ms Davis and Mr Batani had to embark on a journey of unraveling the complex and sometimes subtle web of factors that would contribute to successful curriculum integration. The process of change was slow and challenging, but they believed that the long-term benefits of helping the students to become more independent learners were worth the effort. The results of the final project show that they may just have succeeded.

Focus Questions

1. In many ways the changes that occurred over the duration of the study could be characterized as a shift from assumptions to expectations. What assumptions did the teachers appear to make in the early stages of the projects? What assumptions can we infer on the part of the students? How did the expectations of teachers and students change by the end of the study?
2. What qualities or attributes did the teachers possess that helped them in their quest to achieve better outcomes for their students?
3. The process of interviewing the students appears to have contributed to the successful outcomes of the projects. How could this be adapted and applied in the normal classroom setting without the presence of a researcher?

Suggestions for Further Reading

Stinson, K., Harkness, S. S., Meyer, H., & Stallworth, J. (2009). 'Mathematics and science integration: Models and characterizations'. *School Science and Mathematics*, 109(3), 153–161.

The authors of this study provide examples of integration in mathematics and science teaching in grades 5 to 8 and examine how these teachers characterized integrated teaching. The authors point out that teachers need a clear understanding of what mathematics and science integration is, and is not. Teachers must have a strong sense of what it means to make integration happen, because the potential gains from integration (i.e., time savings, improving student achievement, improving student interest or motivation) are dependent on a common understanding of what integration means.

Pendergast, D. (2006). 'Fast-tracking middle schooling reform: A model for sustainability'. *Australian Journal of Middle Schooling*, 6(2), 13–18.

This article reports on a project to investigate which practices, processes, and strategies best promote the development of lifelong learning in the middle years of schooling. Literature from three broad areas, lifelong learning, middle schooling, and school reform, is combined to inform a model for educational renewal. The importance of learner- and learning-focused programs, student engagement, and academic outcomes are highlighted.

References

Ben-Yehuda, M., Lavay, I., Linchevski, L., & Sfard, A. (2005). 'Doing wrong with words: What bars students' access to arithmetical discourses'. *Journal for Research in Mathematics Education*, 36(3), 176–247.

Caskey, M. (2002). 'A lingering question for middle schools: What is the fate of integrated curriculum?' *Childhood Education*, 78(2), 11–18.

Doerr, H. (1996). 'Integrating the study of trigonometry, vectors and force through modeling'. *School Science and Mathematics*, 96(8), 407–418.

Perkins, D. N. (1991). 'Educating for insight'. *Educational Leadership*, 49(2), 4–8.

Schoenfeld, A. H. (1987). 'What's all the fuss about metacognition?' In A. H. Schoenfeld (Ed.), *Cognitive science and mathematics education* (pp. 189–215). Hillsdale, NJ: Lawrence Erlbaum Associates.

Vygotsky, L. (1986). *Thought and language*. Cambridge, MA: MIT Press.

3

FOCUS ON PROBLEM-SOLVING

Modeling an Ice Hockey Rink at Greenwich Public School

Sheryl MacMath

Introduction

Curriculum integration can take many forms. In one classroom, an integrated unit may focus on a specific problem (e.g., what are the most significant things a student can do to reduce his or her eco-footprint?). In another classroom, an integrated unit may focus on a culminating project (e.g., building a bridge that can support a certain amount of weight). Both of these examples involve a significant end goal. Curriculum integration can also take the form of a thematic unit. Thematic integration typically involves a number of smaller tasks related to the theme itself—there may be a final culminating project or there may not. However, each smaller task will usually involve an amalgamation of disciplines. This chapter details what happened when a teacher, Ms Perry, and her grade 6 students engaged in a thematic ice hockey unit with a culminating project: the building of a scale model ice rink.

To gather information on how student learning developed throughout this hockey unit, a team of researchers observed Ms Perry and her class when they were working on their unit. The research team conducted interviews with Ms Perry and ten pre-selected students before, during, and after the unit. While observing the class during the hockey unit, we closely tracked the conversations and group work of these ten students, monitored the overall classroom environment, took pictures, coded student behavior and activity, and worked with students individually, in their groups, and with them as a whole class. Based on the interviews and observations a comprehensive look at an integrated hockey unit in Greenwich Public School became possible. The research team visited Greenwich 25 times with a total observation time of 33 hours.

Greenwich Public School

Greenwich is an elementary public school in a large urban area. There are just over 375 students enrolled, ranging from kindergarten to grade 6 (approximately 5 to 12 years of age). Typical of most elementary schools, students remain primarily with one teacher in a single classroom for each school year. The weekly schedule for the average elementary classroom is usually portioned into subject blocks. As mandated by the state education department, required curricula include language arts (reading, writing, speaking, listening), mathematics, social studies, science, physical education, and fine arts (visual arts, music, dance, and drama). The average day in an elementary school may start with language arts and continue with mathematics after a recess break. The afternoon would usually be dedicated to social studies or science. These may alternate with fine arts while physical education is usually allocated in 40-minute blocks throughout the week depending on the schedule of the gymnasium.

In contrast to the normal scheduling of subject areas, when working on her thematic unit, Ms Perry scheduled blocks of time dedicated to ice hockey. Throughout their five-week unit, students worked on the hockey unit for three or four days each week. Sometimes it would be for the afternoon; sometimes it would be instead of language arts or mathematics. Regardless, the thematic unit was scheduled in place of a variety of subject areas. As a result, for example, instead of getting out their mathematics books and preparing to multiply or divide, students did not know in advance what specific discipline knowledge or skills would be required for the hockey unit.

Ms Perry had designed her ice hockey unit a few years previously and implemented it each year with her grade 6 classes. Hockey was a long-time passion for Ms Perry that she hoped she could share with her students. Ms Perry also shared with us that she found it necessary to include integration in her classrooms to provide fewer major assignments—having projects that provided marks for a multitude of subject areas reduced time pressures when preparing for assessment and report cards. Reflecting on the previous hockey units she had taught, Ms Perry also emphasized the value of motivation: "The boys … catches their interest … girls, they take a little while to warm up to it, but when they see me as a female teacher, so passionate about the sport, they kind of take my lead." The unit itself was based on the premise that a local hockey team was being newly created. Students worked in small groups (of four to five students) to design their team name and logo, design their stadium, create their schedule, and manage their players. This year, Ms Perry added a new task: groups were also required to build a model ice rink to scale with a working light system. As displayed in Table 3.1, there was a variety of activities with each activity requiring various degrees of integration of subjects.

For example, while the calculation of player statistics focused only on mathematics, the ice rink required students to use knowledge and skills from three subject areas (mathematics, science, and art). Students moved through these activities in a

TABLE 3.1 Tasks and Learning Outcomes for Ms Perry's Hockey Unit

Task	Learning Outcomes
Design your team name and logo (representative of your city)	• Use a variety of resources and tools to gather, process, and communicate information (SS) • Produce two-dimensional works of art that communicate a range of ideas for specific purposes and to specific audiences (A)
Using a pre-determined budget, determine player salaries	• Make connections among mathematical concepts and procedures, and relate mathematical ideas to situations or phenomena drawn from other contexts (M) • Determine the theoretical probability of an outcome in a probability experiment, and use it to predict the frequency of the outcome (M)
Create a schedule of games for your season (including a budget for travel)	• Use a variety of representations of mathematical ideas and make connections among them (M) • Demonstrate understanding of relationships between fractions, decimals, and percents (M) • Use a variety of resources and tools to gather, process, and communicate information (SS)
Graph and present your player stats	• Collect and organize discrete or continuous primary data and secondary data and display the data using charts and graphs, including continuous line graphs (M) • Read, describe, and interpret data, and explain relationships between sets of data (M)
Create a brochure advertising your city	• Read and demonstrate an understanding of a variety of literary, graphic, and informational texts, using a range of strategies to construct meaning (LA) • Recognize a variety of text forms, text features, and stylistic elements and demonstrate understanding of how they help communicate meaning (LA) • Use a variety of resources and tools to gather, process, and communicate information (SS) • Use editing, proofreading, and publishing skills and strategies, and knowledge of language conventions, to correct errors, refine expression, and present their work effectively (LA) • Produce two-dimensional works of art that communicate a range of ideas for specific purposes and to specific audiences (A)
Write a mini biography for each player	• Use a variety of resources and tools to gather, process, and communicate information (SS) • Communicate mathematical thinking orally and visually (M) • Make judgments and draw conclusions about ideas in texts and cite stated or implied evidence from the text to support their view (LA) • Read and demonstrate an understanding of a variety of literary, graphic, and informational texts, using a range of strategies to construct meaning (LA) • Generate, gather, and organize ideas and information to write for an intended purpose and audience (LA)

TABLE 3.1 (continued)

Task	Learning Outcomes
	• Draft and revise writing, using a variety of informational, literary, and graphic forms and stylistic elements appropriate for the purpose and audience (LA) • Use editing, proofreading, and publishing skills and strategies, and knowledge of language conventions, to correct errors, refine expression, and present their work effectively (LA)
Build a scale model of your ice rink with working lights and a buzzer	• Estimate, measure, and record quantities, using the metric measurement system (M) • Classify and construct polygons and angles (M) • Describe and represent relationships in growing and shrinking patterns (M) • Design and construct a variety of electrical circuits and investigate ways in which electrical energy is transformed into other forms of energy (SC) • Produce three-dimensional works of art that communicate a range of ideas for specific purposes and to specific audiences (A)

Note: SS = Social Studies; M = Mathematics; LA = Language Arts; SC = Science; A = Art

linear fashion with each group maintaining a folder or portfolio to keep track of their work as well as each draft. While many of the tasks required students to learn new ideas (e.g., creating a series circuit, designing a brochure), two activities in particular offered students the opportunity to apply knowledge and skills they had already learned. The difficulties that arose during these activities (creating a team schedule and building an ice rink) demonstrated how gaps in student understanding can remain hidden and how the context of integration can provide an illumination of those gaps.

A Team Schedule: How Many Games Need to be Played at Home?

Prior to the hockey unit, students had already completed numerous mathematics classes, including many worksheets, textbook activities, and various word problems based on percents, fractions, and decimals. Ms Perry had tested students on their ability to use percentages, fractions, and decimals, and most of the class were successful. As a result, Ms Perry expected that students would use their mathematical knowledge to make a team schedule. In their groups, students were instructed to complete the following task:

Your group needs to create a game schedule for your team. This season, your team will play 80 games with 30% of these games being played at home. Using the calendar, the attached lists of teams, and an atlas, determine your game schedule. You cannot have more than two back-to-back games at home.

Ms Perry believed students would quickly determine that 24 games needed to be played at home, and expected students to spend the majority of their time calculating distances between cities throughout the United States and Canada to determine the best travel schedule for their team. Even though students were excited by the task, and eager to start planning their routes, they knew they first needed to know how many games were to be played at home. However, what was 30% of 80?

Ms Perry was in shock. For a full hour, every student in the class was worked hard to calculate 30% of 80. Students tried adding 30 and 80 together; 110 was too big a number so they dropped the zero; 11 + 11 + 11 ... that did not lead anywhere. Students tried something different. They divided 80 by 30, but that gave a decimal; you could not have part of a game! Within the first ten minutes, most groups realized that 30% was just a little more than half of 50%. Students realized that 50% of 80 was 40 and half of 40 was 20. Therefore, the answer needed to be a little bit more than 20. Students had calculated what a reasonable answer would be, but they had no idea how to calculate the correct answer. As Ms Perry walked around the room, she noted how creative students were in their problem-solving; however, she was surprised that students did not know the correct procedure for solving what she felt was a simple part of the task. We listened as Ms Perry repeatedly exclaimed, "But they know how to do this!"

Ms Perry began asking leading questions of each group. "What is 30% as a decimal?" Students had not considered converting a percent to a decimal. Students quickly determined that 30% was the same as 0.3; they then began trying a variety of mathematical combinations. Students focused on trying to divide 80 by 0.3 and then repeatedly adding. Then they would switch to subtraction; most students avoided multiplication. When asked why, students shared that they did not think that multiplication would work because multiplication just made the number bigger and they knew that the answer could not be more than 80. While students' estimation and reasoning skills were strong, they had a misconception regarding multiplication as well as a gap in knowledge.

Eventually, Ms Perry stopped the class and provided direct instruction on how to tabulate 30% of 80. As soon as Ms Perry began the procedure, most of the students in the room caught on immediately and quickly figured out that 24 was the answer. As soon as they had the answer, they moved on to the more exciting task of making travel plans. The gap in knowledge revolved around knowing when to use a specific mathematical procedure—in this case, multiplying 0.3 by 80 to give 24. Ms Perry also demonstrated how students could easily figure out 10% (by simply moving the decimal over one column to the left) and then multiply that answer by 3 ($8 \times 3 = 24$). For both strategies, students demonstrated that they knew the strategy and could perform it, but they did not realize that this was needed for this problem. Ms Perry was surprised by this, noting that students had already completed word problems similar to the scheduling problem. The big difference was context.

Within the mathematics class, when working through the unit on percentages, fractions, and decimals, students had encountered numerous problems. During the unit they were introduced to several procedures, modeled the examples provided, and were successful. There was little question regarding which procedure to use; the unit focused on moving between percentages, fractions, and decimals so, of course, this was a strategy that students readily used. However, in the context of the hockey unit, with no model questions or unit examples, students had little guidance regarding which mathematical procedures were appropriate. Consequently, this gap in knowing when to use a specific procedure was illuminated. By changing the context, Ms Perry gained a better understanding of her students' ability (or, in this case, inability) to apply their mathematical knowledge to new situations.

Building a Rink to Scale: Do We Have to Measure Everything?

With sheets of Styrofoam, popsicle sticks, glue guns, and paint sitting in front of them, Ms Perry's grade 6 class could hardly contain their excitement; they could not wait to begin construction on their very own ice rink. Students raced to the computers to determine the dimensions of an actual ice rink: 61 m by 26 m. In previous mathematics units, students had worked with ratios. They had enlarged and shrunk drawings by specific ratios (e.g., draw this square three times larger than shown here). This was the first time students had been introduced to the term, "to scale." Ms Perry explained to students that, to make something to scale, you needed to make it "exactly" like the real rinks, only smaller. She stressed that all measurements needed to be "shrunk" by the same amount. How groups were going to shrink their rink and construct it using Styrofoam was left up to students. Again, Ms Perry stood back and let students wrestle with the problem. All groups came up with a similar plan: centimeters for meters, this was the secret.

Most groups focused on the length of the rink. In reality, ice rinks are 61 m long. By attaching two Styrofoam sheets together (46 cm and 15 cm) students were able to make a model rink that was 61 cm long. By substituting centimeters for meters, most students felt they were constructing their model to scale. However, their scaling work then stopped. Rather than determining the proper width of their rink, students left the Styrofoam as it was. For these groups, their rinks were square-like and "looked right." In contrast, one group measured both the length and the width, making their model 61 cm by 26 cm. Their rink was long and skinny in comparison. When students shared their rinks part way through construction, other groups commented on the long, skinny rink. They felt that it did not look quite right. However, when students printed off a picture of an ice rink on the computer, they were surprised to see that rinks were long and skinny as opposed to the square-like images students had in their heads. At this point, students in the class realized the importance of measuring all dimensions when drawing or constructing objects to scale.

Again, a misconception had been brought to light. Ms Perry was surprised that most of her students were not concerned with shrinking all measurements. In their previous work with ratios, students had worked with many drawings where they had to either reduce or enlarge the image. In each of those drawings, students altered all measurements. Why did they not do this when constructing the ice rink? The answer to this question relates to the type of re-drawings that students were working with. During previous mathematics classes, students had worked on sheets with an image and then had to redraw it, larger or smaller, often right beside the original image. Students would look at the image they had drawn and evaluated whether it looked right. If they had not altered all measurements, their picture did not look right. The focus here was on looking right … if it looked right, it was correct. In contrast, when working on the task of building a model ice rink, students were not originally working with a picture of a rink on paper. Instead, they were working with the image they had in their heads. Even though they had not calculated the correct width, it still looked right. Therefore, students believed their rink was to scale. The misconception involved an understanding of why an image or object is to scale. Having an object look right does not make it to scale—ensuring all the measurements are adjusted using the same ratio makes it to scale.

At the conclusion of the unit, we interviewed a number of the students. During these interviews a second misconception emerged. When we asked students to draw a number of real objects to scale, students assumed that this meant they had to alter the measurement units (for example, meters to centimeters). We specifically asked students to draw a desk that was eight times smaller than the desk they were sitting at. When they measured the desk, it was in centimeters. Did that mean they had to use millimeters? That was too small to calculate. Students were stumped. They knew that all measurements had to be recalculated, but they had assumed that scaling meant changing the measurement unit. Students had not understood why altering from meters to centimeters, for the ice rink task, was considered a ratio. Students had not realized that switching from meters to centimeters meant that they had reduced the ice rink 100 times (as there are 100 centimeters in a meter). As a result, when students were asked to reduce the picture of the desk eight times, they were unable to apply their knowledge of ratios to the task. When we demonstrated to students how to shrink a measurement eight times (e.g., dividing the length of the desk, which was 56 cm, by 8 to get 7 cm), students remembered working with ratios and making these calculations. Again, this illustrated a lack of knowledge regarding when to apply a specific procedure.

While students knew the procedure for determining ratios, they did not realize that this approach was needed when working with the scaling of the ice rink. When students had originally been taught how to calculate ratios when shrinking and enlarging images, they were not taught to identify when they needed to employ these strategies. By providing a unique context, the integrated task of building an ice rink to scale illuminated a gap in students' knowledge of ratios. A new task, with a context separate from the modeled questions present in the usual mathematics class, illuminated whether or not students knew when to employ strategies associated with ratios.

When we asked Ms Perry to reflect on the ice rink activity, we specifically asked her if she would keep it, change it, or not use the rink activity again. She answered that, "I would [do the rink again], yeah because they learned a lot from working as a team … to get their dimensions correct." Given the enormous amount of time students spent on building attachments to their rink (which were not to scale), Ms Perry revealed that she already had a different plan for the implementation of the rink assignment with her next class:

> I would probably focus a lot more on [ratios]. I think I would reduce it. I wouldn't even have the option of stands and boxes. I think we could throw that in after they've done the objectives … okay, now go crazy … next year, you need to get the scale done.

In addition, Ms Perry believed that it would help students remain more focused on the appropriate objectives by giving students the project assessment rubric so that they knew what aspects of their project were the most important.

Discussion

Outside of the classroom, students encounter an infinite number of problems. For example, when trying to buy food at the grocery store, students need to be able to determine if they have enough money to pay for their purchase. Part of the cost of that purchase may be an added sales tax (for example, 6%). To determine if they have enough money, students need to know the procedure for calculating the percentage of a total (for example, $6\% \times \$8 = 0.06 \times 8 = 0.48$ making the total cost $8.48). To solve this problem, knowing the correct sequence of steps is not enough. Students need to know which sequence of steps is appropriate. Why is it helpful to think of a percentage as a decimal? Why is multiplication the best operation to perform? Knowing the answers to these questions will help students decide when to use a specific strategy. The challenge that can arise in the subject-specific classroom involves the clues provided to students. When working in the social studies classroom, the science lab, or the mathematics classroom, students are aware of the strategies that are appropriate to that subject area, because they are given strong clues regarding the type of strategies that will be required of them. The number of clues then increases when students are working on a specific unit of study (e.g., addition of fractions, lifecycle of a butterfly, guided reading). However, the use of an integrated unit removes these clues because it changes the subject-based context. To be successful, students must have the knowledge regarding when and why certain strategies should be used. If they do not, this gap in knowledge becomes evident.

As the two examples outlined in this chapter illustrate, by providing students with the opportunity to work through tasks independent of specific subject areas, gaps in student understanding were illuminated. In both cases, these gaps were unexpected. As students began to work through the different tasks outlined in this integrated hockey unit, the lack of contextual clues left students trying a variety of strategies hoping to come across an answer that made sense. It was as though they were blindly searching for a magic sequence of steps that would result in the correct answer. The challenges they experienced when working through the game schedule and the construction of an ice rink called students' knowledge into question. What counts as understanding? What counts as learning?

If students are unable to determine when and why a strategy is appropriate, they lack the ability to transfer knowledge to new situations. Ross and Hogaboam-Gray (1998) theorize that the broad context associated with integrated units enables students to transfer knowledge to a larger number of situations. By not limiting students to a subject-specific problem, students can apply what they learn in an integrated setting to a greater variety of circumstances. Curriculum integration, by removing the usual contextual clues of the subject-specific classroom, can provide an invaluable opportunity for teachers to assess student understanding and work towards enabling students to apply knowledge in settings beyond the school classroom.

Focus Questions

1. The students' inability to scale their ice rink correctly provided Ms Perry with a challenge. Should she allow students to spend two weeks working on an ice rink that is not to scale or should she stop them, teach about ratios, and then have them complete their rinks to scale? What options exist for Ms Perry? What would you do?
2. Our observations of this unit illustrated how the real-life tasks of the integrated context illuminated student misconceptions. In this case, a lack of conceptual knowledge regarding percentages and ratios was revealed. When building your own integrated unit, once you have identified your objectives, try to identify a number of possible misconceptions that students may have.
3. The hockey unit was scheduled after Ms Perry had formally taught what she thought to be the prerequisite knowledge and skills in a disciplinary setting. For example, she taught electricity in science, percentages and ratios in mathematics prior to the hockey unit. How would this unit have been different if these disciplinary knowledge and skills were taught within the unit rather than prior to the unit? What effects do you think this would have had on student learning? On student motivation? What would you do?

Suggestions for Further Reading

Clement, L. & Sowder, J. (2003). 'Making connections within, among, and between unifying ideas in mathematics'. In S. A. McGraw (Ed.), *Integrated mathematics: Choices and challenges* (pp. 59–72). Reston, VA: The National Council of Teachers of Mathematics, Inc.

In this chapter, the authors discuss how many students in mathematics classes use immature, non-sense methods to solve problems. The authors caution that these strategies inhibit students' ability to solve problems, and, more importantly, inhibit their ability to understand the nature of numbers and operations.

Doerr, H. (1996). 'Integrating the study of trigonometry, vectors and force through modeling'. *School Science and Mathematics*, 96(8), 407–418.

This case study examines the development of students' understanding of the motion of an object down an inclined plane using a modeling-based approach. A key strategy was to encourage students to discuss, to make conjectures, and to construct arguments based on evidence. The author argues for more time to be spent using this strategy and less time on drill, practice, memorization, and lecturing.

Reference

Ross, J. A. & Hogaboam-Gray, A. (1998). 'Integrating mathematics, science, and technology: Effects on students'. *International Journal of Science Education*, 20(9), 1119–1135.

4

FOCUS ON ENGINEERING
Bridge Building at Southern High School

Grady Venville

Introduction

Science and mathematics teachers often protest that they don't know how to respond when their students ask, "Why are we learning this?" School science and mathematics subjects are often taught in abstract ways and concepts are not applied to real-world problems or issues. Rather, students learn to problem-solve with algorithms; they learn formulae by heart, internalize methodological approaches, or memorize definitions and explanations without true understanding. Examination-driven curricula support this approach to teaching because it is possible for students to "cram for the exam" and do well without developing in-depth knowledge.

The irony of this non-applied approach to the teaching of science and mathematics is that the fundamental purpose of school is to prepare students to become educated, knowledgeable, and just adults. However, employers frequently report having to educate employees how to "function in the real world" when they have left school (Johnson, 2004). Is it possible for education to be structured in such a way that it provides students with experiences that are genuine, practical, and useful in the real world? Rogers (1997) argued that subjects shouldn't be considered the only way to structure school curricula, and that the professions, such as dentistry, law, and nursing, provide meaningful knowledge and standards that could be used to guide teaching and learning. As an example, Rogers explained that the profession of architecture provides issues that can be addressed in project-based curricula and, further, it provides a set of standards by which students' work can be evaluated.

The challenge for teachers when taking on integrated approaches to curriculum, as would be expected when addressing real-world problems, is that the students are still in school and it is important that they do meet the standards of state or national curriculum documents. Such curriculum documents are almost always structured and written in the form of disciplines. Further, associated examinations focus on discipline-based knowledge. Integrated, real-world projects restrict the content knowledge to that relevant to the problem, and often the teaching and learning of the conceptual aspects of the project are treated in superficial ways (Lederman & Niess, 1997).

The case story presented in this chapter reports our investigation of the implementation of a bridge-building project with grade 9 students (aged 13 to 14 years) at Southern High School. In particular, we were interested to see the degree to which students were able to achieve the engineering criteria that the teacher set for the assessment of the project, and, in addition, the degree to which students achieved conceptual standards that we might expect them to meet in a more traditional, science-based approach to curriculum. In other words, could the students learn both engineering principles, for example, about structures and aesthetics, and science concepts, for example, about forces, when participating in a real-world, engineering project about bridges?

Background

The students in this grade 9 class investigated, designed, and constructed a model bridge. During the teaching and learning process, they examined some theory about the forces involved in putting a load on a bridge. Compared with mainstream science classes, however, the bridge project required the students to develop much more applied and practical understandings. Understanding of forces has been documented as a challenging area for students in middle and high school, and students tend to have a number of related misconceptions (Driver, Squires, Rushworth, & Wood-Robinson, 1994). For example, a common misconception is the view that a table does not exert an upward force on a book resting on the table. One of the biggest challenges for teachers is that students tend to associate forces with movement, not recognizing the passive forces involved in equilibrium situations, like the book on the table. That is, they think that forces are acting only when something is moving and that, if there is no movement, there are no forces acting. This way of viewing the world seems to be a result of learners thinking of forces as a property of a single object, rather than as a feature of interaction between two objects (Driver et al., 1994).

One way of addressing the problems and misconceptions students have with learning about forces is to use structured scaffolding. For example, it is possible to talk about the forces acting when a book is resting on a spring. In this situation, students are more able to recognize equal forces acting in opposite directions between the book and the spring. The instructor then moves on to a flexible table, and then returns to the scenario of a book on a standard table. In this way, learners are more likely to develop the understanding that the table exerts an upward force on the book (Brown & Clement, 1989).

Case Story: The Bridge Construction Project

Ms O'Reilly was the teacher at Southern High School who developed this engineering course, incorporating science and mathematics principles and practices within engineering projects. The engineering course is a grade 9 optional course comprising two 1-hour lessons per week. Ms O'Reilly explained that a wide range of students participate in the course, including some who are academically talented, as well as students who are skilled in practical ways. Ms O'Reilly said that, regardless of their varied skills and talents, most of the students are "interested and motivated." The class involved in this case story comprised 15 male students. The students were involved in several projects throughout the school year, one of which was the bridge project.

The students worked in groups of three to role play a bridge construction company over a period of five weeks. The students were informed that another company had gone bankrupt, and that one of their bridges was unfinished. Their task was to complete the unfinished bridge by applying their newly developed knowledge of structures. The role play construction companies were required to produce strong, aesthetically pleasing bridges, but, at the same time, they had to minimize the completion costs. Design criteria were outlined in the design brief and included a span of 750 mm with no support or legs. The bridge also had to have a support capacity of two cartons of soft drink cans. Other criteria were that the bridge was to have a maximum vertical deflection of 25 mm under full load; that it was to be constructed from materials bought from an official supply list, and with tools available in the school workshop. The cost of completing the bridge was to be under 150 "bridge bucks" (play money supplied by the teacher).

In the first two weeks of the course, the students completed several investigations into structures, beams, bending, joints, and jointing. They also were introduced to types of forces, including static and dynamic forces, the history of bridges, and bridge types. Students were required to plan the bridge-building process and then design, manufacture, and evaluate their own bridge. The final aspect of the project was the class evaluation where prizes were awarded to the bridge with the best strength-to-weight ratio, the most aesthetically pleasing bridge, and the cheapest bridge that met all the design criteria. A prize was also awarded to the bridge construction company who submitted the best written assignment.

Engineering Learning

Five bridge-building companies were formed in the class and each company was able to construct a bridge within the design specifications ready for the testing day at the end of the project. Table 4.1 provides an overview of each company, the student members, the bridge design, how much the bridge deflected under full load, the budget used by the company to construct its bridge, the approach the company took to aesthetics, the main problems the company faced, and any prizes won.

Bridge Design

Four of the five companies constructed a deck bridge with plywood and "I" beams to strengthen the structure. Company A added suspension to their deck bridge and Company E designed a double-layer deck bridge with "I" beams between the first layer of plywood and Styrofoam between the second layer. Most of the students in these companies tested various structures and did calculations to see which structures would be the strongest. For example, the students in Company C, Adam, Daniel, and JJ, tested several structures in class and worked out the strongest. Adam felt it was really important to learn about the different structures, beams, and triangles, for example, that were alternatives to "just putting planks on planks." They then did simple calculations to estimate the costs of the various structures and, as a result, they decided to use "I" beams. This company's bridge did not deflect when fully loaded with soft drink cans and they only spent $88. As a result, Company C won the prize for the cheapest bridge to meet the design criteria (see Table 4.1).

Lawrence worked with David and Cain in Company E. The students made a double-layered deck bridge with the bottom layer consisting of "I" beams between two pieces of plywood and the top layer comprising a layer of Styrofoam with a third piece of plywood. The group members found that the "I" beams were the strongest structure from their testing, and Lawrence explained how they got their idea for the two-layered bridge:

> It didn't take long to make. We "stole" the design from two people's bridges piled on top of each other. We saw them on the desk while people were putting away their stuff and that's where we got the idea.

The only company not to use a deck design for their bridge was Company D. Steven worked with Craig and Robert to construct their suspension bridge, which he described as

> a suspension-type structure. So we had Styrofoam and plywood in it with bits of string coming down from poles and [from the top] going underneath and supporting it with bits of pipe. We glued all that together so it was like, if it bent, we tried to make it so that the suspension would pull the middle up.

On the day of testing, Company D's bridge deflected slightly under full load, but not more than the 25 mm allowed in the design specifications. Company D won the prize for the bridge with the best strength-to-weight ratio because their bridge was strong enough to hold the required load but, due to their use of Styrofoam between two pieces of plywood, it was much lighter than the companies that had included "I" beams made of plywood for strength.

TABLE 4.1 An Overview of Each Company's Approach to Bridge Construction

Company	Students	Bridge design	Deflection under full load	Budget	Aesthetics	Main problems	Prizes
A	Gavin, Colby, Masahito	• Deck bridge with suspension • Plywood • "I" beams	<25 mm	$100	• Suspension with doweling and string • Spray paint • Covered "I" beams with corrugated cardboard	• Fixing top deck	Best-looking bridge
B	James, Andrew, Hossain	• Deck bridge • Plywood • "I" beams	<25 mm	$150	• None	• Budget	None
C	Adam, Daniel, JJ	• Deck bridge • Plywood • "I" beams	No deflection	$88	• String rails • Colored deck with charcoal	• Calculating amount of string needed	Cheapest bridge to meet design criteria
D	Steven, Craig, Robert	• Suspension • Plywood and Styrofoam	<25 mm	$122	• Suspension with colored string • Colored deck with charcoal	• Hot glue dissolved Styrofoam • Strength	Best strength to weight ratio
E	Lawrence, David, Cain	• Double layer deck bridge • Plywood • "I" beams • Styrofoam	No deflection	$158	• Painted bridge • Double layered construction	• Budget • Adhering "I" beams with PVA glue	None

Problem-Solving within a Budget

Each of the companies faced problems during the bridge construction process that required complex problem-solving within a limited budget. For example, students in Company D, who constructed the suspension bridge, became concerned during one lesson that their bridge would not be strong enough to hold the required load. Steven, Craig, and Robert debated whether to buy two more pieces of plywood so they could add some "I" beams to the base of their bridge. However, they did not have enough money to buy more plywood. Steven discussed the situation with Adam, a student from Company C, and asked him how much plywood they used for their bridge. James, from Company B, suggested that Steven and Craig also needed to think about the glue, advising that they wouldn't have enough glue left to attach all the "I" beams they were planning. After these consultations with other companies, Steven, Craig, and Robert decided not to add the "I" beams and they completed the bridge-building task with several dollars still in their budget. Steven was pleased with the end result because, "we had spare money in case we needed it and it worked out to being less money ... than what we thought we were going to [spend]."

The main problem for Company E was that, by the end of the project, they had spent eight dollars more than they were allocated (see Table 4.1). Lawrence said they solved the problem by borrowing the extra money from Ms O'Reilly. Lawrence admitted that the group did not work out how much money their design would cost before they started constructing it. "We sort of made it up as we went along." The double-layer deck bridge they constructed was very strong, and didn't deflect at all during testing, but the weight of the bridge was comparatively high, so this group didn't win any prizes. During the construction of the bridge, Company E also had difficulty adhering the "I" beams with PVA glue. Lawrence said that one important thing he learnt during the project was that, "PVA glue doesn't work very well on the plywood for the 'I' beams; the glue gun [hot glue] was good with the 'I' beams."

James said he was pleased with the bridge they built for Company B, but if he did it again he would try to spend less money because they had used all their allocated money and didn't have any left for improving the aesthetics of their bridge (see Table 4.1). James explained that, initially, they had put in 10 "I" beams but they thought it wouldn't be strong enough, so they put in another 10 to make a total of 20. James said that other groups had put in fewer beams and they weren't as strong as Company B's bridge, but they were strong enough to hold the required load. James admitted that budgeting was the biggest problem for his company, and that the most important thing he learned from the project was to "save your money ... make it strong, but don't spend as much."

Aesthetics

Each of the companies gave consideration to the aesthetics of its bridge construction, with the exception of Company B who ran out of money and couldn't afford any aesthetic improvements (see Table 4.1). Company C students decorated their bridge with string rails along the edges of the deck, and colored it with charcoal. Company D colored the string on their suspension bridge with pens, and the deck with charcoal so that it looked like an authentic road. Company E students painted their bridge, but Lawrence admitted, "that's the only creativity we used."

In comparison with the other companies, Company A gave more thought to the issue of the aesthetics of their bridge. Gavin, Colby, and Masahito spent considerable time during one lesson debating whether to spend more money strengthening their bridge, or to make it more aesthetically pleasing. They also said they were wondering whether to do this at all, because the bridge was already strong, and they could get a prize for spending less money. They decided to add suspension, "because it will add some strength and it doesn't cost very much." Gavin and Colby decorated their bridge with dowel rods and string, and Masahito spray-painted it, and covered the ends so that the internal "I" beams were not visible. The students then covered the bottom of their bridge with corrugated cardboard "to make it look good." During testing, when two cartons of soft drink cans were placed on top of Company A's bridge, it deflected slightly, but not more than the 25 mm limit. This company won the prize for the best-looking bridge. The teacher from the English department who did the judging on this aspect of the bridge project said that the suspension presented pleasing curves, the bridge had slim, clean lines, and the corrugated cardboard was interesting.

Applying Mathematics and Science Knowledge

Most of the students in each of the companies recognized the utility of mathematics knowledge for aspects of the bridge-building project. For example, for Company A, adding the suspension for strength and looks caused some problems that could be informed by mathematics. Gavin had to work out the cost of the string for the suspension they were considering adding to their deck bridge. He wanted 3000 mm of string and he knew that the string cost $1 for every 300 mm. He had difficulty doing the proportional problem to work out how much he had to pay. Ms O'Reilly helped him work out that the string would cost $10 by doing the cross-multiplication on the whiteboard. When Gavin saw the calculation on the board he said, "We do those all the time in math."

Adam, from Company C, found that his mathematics knowledge was useful for "totaling things up, working out measurements, strength, and things like ratios." Similarly, Steven, from Company D, said his mathematics knowledge was useful to measure, especially because they had to allow 2.5 mm where the saw removes the wood. They also had to measure the string, and align through accurate measurement the poles and string on each side of the bridge. Steven also used his mathematical knowledge to add up the cost of all the materials, and work out how much the bridge would cost before they started. Lawrence, from Company E, argued, however, that his mathematics knowledge wasn't very useful during the project because the mathematics involved was "fairly simple."

Students struggled to identify when their knowledge of science was helpful in the bridge construction project. Gavin, from Company A, was one of the few students who recognized that the experimentation they did at the beginning of the project could be considered scientific experimentation. Adam, from Company C, recognized that he was doing "research" when he did the investigations and he said that he liked doing the research, but he did not associate the investigations with science: "I don't think [science was useful] so much for this project, but for some of the other projects," like a Lego racing car project they had conducted earlier in the year. In contrast, James said that, in Company B, they had tested different structures and found that the "I" beams were the strongest, but he had already learned that in a previous woodwork class. James, and most of the other students, didn't recognize the testing process as scientific investigation.

Understanding the Concept of Force

In order to gain insight into whether the project improved students' understanding of the scientific concept of force, we conducted more in-depth interviews with five students, one from each company; Gavin, James, Adam, Steven, and Lawrence. Specifically, we wanted to know if the students recognized that, when a load is on a bridge, there are equal and opposite forces acting between the load and the bridge. All five students recognized that there were forces in action between the load and the bridge, even though there is no suggested movement. This is in contrast with the findings from research discussed earlier, that the majority of students of this age associate forces only with movement.

When questioned more deeply, however, only three of the five students, Adam, Gavin, and Steven, recognized that forces were acting in opposite directions. Moreover, only Adam and Gavin identified some kind of balance between the forces that results in the equilibrium situation of the load on the bridge. For example, Adam, from Company C, said that the kind of forces acting are, "just static forces, hitting one place, just going down ... also shear force here [the sides of the bridge], that's a force down and that's a force up like that." When asked to explain further Adam said, "that's [the bridge's] just holding it [the load] there, so there is an equal push down and up."

Gavin, from Company A, was less confident than Adam and, while he recognized the static forces, he only seemed to work out the issue of forces acting in both directions during the discussion with the interviewer. Gavin said that there were "static" forces because "it's just holding up itself like by the strength of the wood or whatever the materials are." When asked if the forces had any direction, Gavin said that, "mainly it's just, well, the load is pushing down, but it won't go down unless it was going to snap." The researcher asked Gavin why the load stays there and he answered, "because of the strength of the bridge." When asked if there is a force acting up on the load he said, "I suppose, the strength of the bridge would be pushing up ... that would even out until there was a larger load here, then it would overcome this one, which is a set load I suppose, and it would push down and it would break."

The other three students all had some misconceptions about forces. For example, Lawrence, from Company E, said that he didn't think there were any forces acting on the load, but there was on the bridge. The interviewer asked him what forces were acting on the bridge and he replied, "the load." Lawrence's notion that the load was the force acting on the bridge suggests that he saw a force as a property of a single object (the load) rather than an interaction between two objects.

Conclusion

This case story of an engineering project at Southern High School provides considerable information about students' learning. The project outcomes were assessed by the teacher according to engineering criteria of design, budget, and aesthetics. The students were involved in complex problem-solving. They were required to find their own solutions, for example, how to increase the strength of their bridge while keeping costs to a minimum. This engaged the students in thinking about the materials available and their properties because they had to make decisions about their bridge design based on this knowledge. The tests the students conducted at the beginning of the project with beams and structures played a significant part in the decision-making process.

Students were encouraged to be creative, and a prize was awarded for the most aesthetically pleasing bridge. This factor created an alternative dimension to the bridge-building project that complicated the process of problem-solving. The students not only had to find solutions for the problems they encountered, but also the solutions had to be within parameters for strength, cost, and aesthetics. This engaged the students in complex cost–benefit analysis.

The social aspects of engineering were apparent in the bridge construction project. Students within companies worked together to test materials and conferred with each other to make decisions about their bridge. The social aspects of learning also were evident between the companies, for example, when Steven, from Company D, consulted Adam, from Company C, about the materials his group had used.

Mathematics knowledge was applied by all companies of students during the bridge construction process; however, this was acknowledged by some students as low-level mathematics. With regard to science learning, the case story showed that students recognized that forces were in action in a static, "load on a bridge" scenario. There were, however, indications that some of the students held misconceptions, in particular that a force is a property of a single object.

Within the context of this bridge design project, Ms O'Reilly's students were immersed in real-world problem-solving while also applying important scientific and mathematics concepts and principles. As well, they learned skills that are valuable in a bigger, global context, including time management, cost–benefit analysis, and cooperative social skills.

Focus Questions

1. How could the teacher in this case story have improved the students' learning about forces?
2. While the teacher in this case story was a woman, no female students selected this course. Why do you think this might be the situation? Would high school girls benefit from an engineering course like this one as much as boys? How can more girls be encouraged to participate in engineering?
3. Which do you think is more important? Abstract, discipline-based knowledge that will help students pass exams? Or integrated knowledge that will help students in the real world? Why?

Suggestions for Further Reading

Venville, G., Sheffield, R., Rennie, L., & Wallace, J. (2008). 'The writing on the classroom wall: The effect of school context on learning in integrated, community-based science projects'. *Journal of Research in Science Teaching*, 45(8), 857–880.

This article provides an in-depth examination of how the school context impacts on the type of learning that occurs when an integrated project is implemented. The high school in this story focused on student learning of science content knowledge, while the middle school focused on the teaching and learning of values such as social and civic responsibility, and environmental responsibility.

Johnson, J. (2005). 'Practice drives theory: An integrated approach in technological education'. In S. Alsop, L. Bencze, & E. Pedretti (Eds.), *Analysing exemplary science teaching* (pp. 84–95). Maidenhead, UK: Open University Press.

This chapter is an account of an integrated technology education course that challenged students to design and build a model car powered by a mousetrap. It looks at the challenges of implementing the technological design process in the school context, including the issues around open-ended problem-solving, physical manipulation of apparatus, and conceptual integration.

References

Brown, D. E. & Clement, J. (1989). 'Overcoming misconceptions via analogical reasoning: Abstract transfer versus explanatory model construction'. *Instructional Science*, 18(4), 237–261.
Driver, R., Squires, A., Rushworth, P., & Wood-Robinson, V. (1994). *Making sense of secondary science: Research into children's ideas*. London: Routledge.

Johnson, M. D. (2004). 'The newest "reality show": The importance of legitimizing experiential learning with community-based research'. *The American Biology Teacher*, 66(8), 549–553.

Lederman, N. G. & Niess, M. L. (1997). 'Integrated, interdisciplinary, or thematic instruction? Is this a question or questionable semantics?' *School Science and Mathematics*, 97(4), 200–205.

Rogers, B. (1997). 'Informing the shape of the curriculum: New views of knowledge and its representation in schooling'. *Journal of Curriculum Studies*, 29(6), 683–710.

5

FOCUS ON LITERACY

Linking Language and Horticulture at Seaview Community School

Susan Joan Gribble and Léonie Rennie

Introduction

We left the tiny airport at about 2 p.m. on Saturday and, after camping overnight, we arrived at Seaview Community School mid-afternoon on Sunday. It wasn't difficult to locate the school; it was the only cluster of buildings apart from the local gas station and the spread-out community housing. It was surrounded by a strong wire fence, not to keep children in, but to keep cattle out! Seaview is located in the middle of a large beef cattle ranch through which the cattle roam freely. The house next to the school was occupied by Mr Lanyon, the principal, his wife Ms Lanyon, who was the grade 4 to 7 teacher, and their two children, who were the only non-Indigenous children who attended Seaview. A second house was shared by Mr Benson, the high school teacher, and Ms Kelly, who taught the lower elementary children. The cluster of buildings was completed by an amenities block and a small, one-person transportable building currently accommodating a pre-service teacher based at Seaview for her weeks of field experience.

Mr Lanyon and Ms Lanyon greeted us enthusiastically for it wasn't often they received visitors. They showed us over the school, which comprised three classrooms; a faculty room, which doubled as the library, canteen, and general storeroom; a small office; and another small storage room where first aid was available. There was no accommodation in the town and the transportable room was not available, so we settled ourselves for a week's stay in the faculty room. We had a refrigerator and power outlets for our electric kettle and cooker, and just enough space on the floor for our sleeping bags, so by ignoring the large numbers of long-legged spiders in the untidy webs festooned around the open louvered windows that ventilated the room, we could be quite comfortable. It was nice to be woken up on Monday by the chattering birds that began their morning activities rather early.

During our week at Seaview, Joan (the first author) was able to interview all of the attending children, testing their literacy and mathematics skills, and assessing their progress over the year since her previous visit to the school. She also interviewed the teachers, talked to parents, and collected data about school–community relationships. Léonie (the second author) interviewed all of the teachers, observed a number of classes, and had numerous conversations with the children. It was unusual to have two visitors staying at the school and it excited considerable interest in the community. More mothers than usual came in with a playgroup one day, and children wanted to ask us lots of questions about our families and why we were there. During the evenings, we generally kept to ourselves, discussing the day's events and writing our field notes, but we enjoyed a barbecue with all of the school staff on our last evening in the community.

The Students, the School, and the Community

The small community comprised mainly women, children, and older people. Many of the 15- to 30-year-olds had moved into towns where life was more exciting. With a cattle drive coming up soon, some of the young men were returning to the community. The ranch gave the Indigenous community economic viability and the community was regarded as safe and stable, allowing the school to get on with its own work, supporting the children in ways that served the community. As principal, Mr Lanyon's approach was straightforward: "Find out what the community expects, what they think is needed for their children, and get on with it," he said. It felt to us that the school and the community existed in harmony.

Even though some of the older people had never experienced schooling, children talked about their activities and everyone seemed to know what was happening at school. The children were well cared for and encouraged to go to school, but there family responsibility seemed to end. Children were not overtly pushed to achieve. As Mr Lanyon pointed out, "school is something that happens between other things in this community. What is done with family is more important than school." Whereas we might fit our children's activities around their schooling, here, school happens in gaps between other things in children's lives.

School attendance was facilitated by a loud siren sounding in the community at 7 a.m. (time to get ready for school) and 8 a.m. (time to start coming to school), because most households did not have clocks. Another attraction to attend school was the presence of heaters for each classroom. These offered great encouragement to students to leave the campfire at home and they were rarely late on cold mornings. Consequently, the students residing in the community arrived early at school, helped by the siren, with evident energy and enthusiasm.

During our week at the school, an average of about 30 students attended each day. Although the school register held considerably more names than this, many students had irregular attendance, often traveling to other locations with family, sometimes attending other schools, and then returning to the community. The resultant transiency created considerable difficulty for the teachers, particularly in the higher grades, because of interruptions to longer-term projects they might wish to plan in science or social studies. It meant that planning had to be short term, because of uncertainty, week by week, about who would attend and what community activity might disrupt school attendance.

Besides the four non-Indigenous teachers, the school had a full-time Indigenous education worker in the Kindergarten to grade 3 class (4 to 8 years), a part-time gardener, and a cleaner. Because the school was small, there was little flexibility to change class structures. Mr Lanyon's role as principal carried a 0.7 time-fraction teaching load (with 0.3 for administration), so he taught each of the three classes at times so that their regular teacher could have time for their "duties other than teaching." Staff were committed to the school and its students. There was a shared vision and unity of purpose, and everyone knew what was going on. Mr Lanyon insisted that all matters, no matter how small, were discussed so that bigger problems did not arise. His supportive leadership, together with their close living together, created a cohesive, family atmosphere among them.

The Curriculum at Seaview Community School

The teachers and parents worked together to set priorities for the school. The school's development plan was written to reflect the community's expressed goals for the school. They expected that children would be taught to read, write, speak "good English," and learn mathematics. The school purpose statement read:

> The students at Seaview Community School will learn to speak, read, write and do mathematics as well as maintain and learn their language [the local Indigenous dialect]. They learn and use all the skills necessary to be a useful and complete community member [in this community]. At the same time they will be able to recognize the need to have the skills to switch [from their role in their own community] and be a useful and complete community member in the wider community.

Teachers believed that parents did not see the curriculum much beyond reading, writing, and knowing about money, and they were concerned that the content of the curriculum needed to be more than those things. Mr Lanyon explained it this way:

> We need to make some really hard decisions about what we can do and what we can't. ... A curriculum framework for these kids should be about what content they need and changing the timeline for their learning. We must teach them how to learn and access "white" thinking. We shouldn't expect the children to be formally reading before grade 3, and we need to take more of our elementary teaching approaches into the high school

classroom. The curriculum needs to focus on how to be successful and a contributor to this community. The kids need to become confident and skilled to do things for themselves and not rely on a white person to do it for them.

The mainstream state curriculum posed some problems for these teachers, but it also allowed for flexibility. They needed to select and plan from the curriculum to present a program that was meaningful and relevant to the student in the context of his or her environment. Mr Benson explained how they tried to do this.

These kids don't get a reduced curriculum as such. We look at science from an environmental perspective like tourism and its impacts and look at different cultures and religions. They are very interested in this type of work. Vegetables and fruit are hard to get hold of in the community because of transport and costs so the school addresses the nutrition issues through the school program. We get the kids to cost, order, and organize the fresh food in, via the plane, for the cooking program and the canteen. We have started a horticultural program to grow fruit and vegetables. Our cooking program is very helpful in the classrooms and we teach many concepts this way. ... What is really reduced for these students is the time they spend at school and the time they do not use the English language when they are not at school.

The small class sizes, usually about 12 students (although the composition of the 12 changed frequently), meant that students were well known by their teacher and could be worked with, one on one, in developing needed skills. But, as Ms Kelly remarked, "my emphasis on planning is around short-term goals based on the differing levels and needs of the kids. Our big problem is that the kids are not always here and this impacts significantly on planning. We waste a lot of learning time."

The Importance of Language

At Seaview, in common with other Indigenous schools, English is taught as a second language. The children's first language is usually a Kriol that is typically spoken in the community. There is also a traditional Indigenous language, which, even if spoken among adults in the community, is generally not known by the children. Many schools, including Seaview, have a second language program in which one or more of the local elders take classes in the traditional language, usually on a weekly basis. However, such elders usually have other responsibilities in the community, so their attendance is irregular. At Seaview, the Indigenous education worker assisted the teachers to integrate this language into programs to help the students increase awareness and use of the traditional language. The language learning strategies were activity based and usually conducted outdoors. Also, Mr Lanyon tried to encourage caregivers to use the traditional language to talk to the children at home, to facilitate their learning, but it seemed to us that its continued existence was under threat from disuse.

English is the major medium of communication between teachers and the caregivers in the community, and this communication is imbued with concern about ensuring mutual understanding. In the classrooms and school activities the teachers focused on what speakers and listeners did when using the English language. Students were taught to ask questions for clarification and were shown how to use different cultural strategies when using different languages. For example, they were explicitly taught how to engage in the learning process and be proactive in trying to learn. As the secondary teacher, Mr Benson relied heavily on the teaching backgrounds of the elementary teachers to help him make judgments about the development of students' knowledge and skills, and how to go about teaching students to progress from one point to another. At the same time, Mr Benson and Ms Lanyon tried to develop students' deeper understanding of the English language, "the secrets," so to speak. Mr Benson introduced poetry and the students discussed, for example, the use of metaphor in the English language. Ms Lanyon explained her view:

> Language teaching has to link form and purpose so the kids know why they learn English. When content becomes more complex children's usual coping strategies don't work any more, so we need to teach them how to interact and engage in the learning process by learning to use "white people's" skills as well as their own, like questioning, reflection, and having another go at things if mistakes occur. We have to get better with this process because this does not happen well enough yet.

The students were good at questioning, but mostly good at "where" and "what" questions, rather than "why" questions, yet so much of being able to apply learning from school depends on knowing the answers to "why" questions. To these students, why wasn't at all important. Things just happened the way they did.

The Integrated Program at Seaview

As we watched the teachers and students working inside and outside of their classrooms, we could recognize different subjects, particularly in the high school class, but there was considerable integration. There was a concurrent "learn the basics" and "learn for living" approach underpinning the curriculum in the school. There was a strong belief that the curriculum framework needed to be developmental across grade levels with a whole emphasis on immersing students in oral language. In each of the classrooms, every lesson, every day, was treated as a language lesson.

As all of the teachers described to us, everything in the school's program is integrated with language. They try to take every opportunity during class to encourage the development of language, using the other subjects as a medium in which to do this. For example, Ms Lanyon referred to the confusion that arose when reading a book in English where a fireman "put the fire out." She discovered that, in the local language, a person would "light the fire out." The ensuing explanation about fire was not only about language but also extended into a scientific discussion of fire and burning.

Mr Benson taught all subjects to the high school students, which made it relatively easy for him to link the subjects together, compared with high school teachers in larger schools who might teach only one or two subjects. He worked with a relatively regular program. Before recess he concentrated on mathematics and language, focusing on the knowledge and skills of the subjects. Between recess and lunch, he generally did a range of activities—he referred to these as "games"—to do with mathematics and language. This provided opportunities to link the two subjects together, often incidentally with other subjects. For example, at the time of our visit, he had been doing one lesson a week on health. It was centered on drugs, smoking, cannabis, and alcohol, and by the last morning he had developed a big explosion chart on the white board about cannabis and its effects. This had provided students with a lot of opportunity for discussion and writing to practice literacy skills.

After lunch the class worked on "option" subjects. Mr Benson clustered these under three names: technology, art, and photography. He had a project planned to make benches out of local natural wood. This would require a range of mathematical and design skills, as well as science about the nature of wood and technology in making the bench. A keen photographer, Mr Benson planned to use photography for students to record the sequence of their work. This project made explicit links with language because the words used in the woodwork, science, design technology, and photography would become list words in his English classes, and the procedure for building the benches was to be written up as one of the genres in English.

The Horticultural Program

One whole-school program that integrated science, technology, and mathematics, and also language, was based around horticulture. Mr Lanyon led this program as part of his 0.7 teaching load. The outcomes of the horticultural program were visible in the fruits and vegetables growing within the school grounds. The school fence was particularly important to its preservation, because the produce was otherwise in danger of being eaten by the roaming cattle. On one of the days we were at Seaview, the children shared in tasting the first fruit harvested for the season. As the cost of trucking fresh fruit and vegetables to this remote area is prohibitive, the teachers also had vegetable gardens.

Mr Lanyon explained that he prepared activities in the horticultural program that were enjoyable as well as meaningful, and would allow maximum participation by students. He wanted to use these activities to develop independence in the students so they could continue on their own without relying on the teacher's presence. As the climate was ideal for horticulture, Mr Lanyon was hopeful that some of the community members would also begin their own gardens, and he was propagating some trees in a small area in front of the school with a simple horizontal sapling fence to model how such plants could be protected from the cattle.

The younger students worked at horticulture in their class group. The K-3 students were planting out, watering, and mulching tomatoes during our visit. The seedlings had been grown from seeds collected from last year's crop. In an exercise combining language and mathematics, these youngsters were counting aloud the handfuls of organic fertilizer added to the soil in which the tomatoes were being planted.

Older students usually worked individually or in smaller groups on various tasks. During our visit, two loads of topsoil were delivered and these students were just beginning the construction of a new vegetable garden for traditional vegetables to supplement the sweet potatoes in the old garden, which had become too shady. The students had measured out the new bed and were spreading the topsoil, removing roots, and digging in compost prepared earlier. Mr Lanyon planned to develop the new garden using permaculture, with cattle dung substituting for artificial fertilizer, and controlling the weeds with mulch. However, to enable the achievement of short-term goals for the students, some fertilizer was being used to get early growth.

The older students kept a journal or log to write about their activities in the horticultural program, and this was an important contribution to their language development. Depending on the weather and the season, activities varied. Sometimes it was more important to do the physical work than the written work. During the wet season, it was often more comfortable to be working indoors, writing and drawing about the activities. Thus, seasonal conditions, the rate of plant growth, and what was ripening when, meant that the horticultural program allowed geography to be linked with science, technology, mathematics, and language activities. It was an excellent example of curriculum integration.

Discussion

We left Seaview somewhat sadly. We had been welcomed, we enjoyed our stay, and we had been fondly farewelled. We had been uplifted by the school atmosphere of harmony, the enthusiasm of the students attending school, and the dedication of the teachers. But there was a sense of concern as well. What we had seen suggested that, although Seaview was what might be considered an excellent school, the children's learning and skills were not on par with what might be expected of children their age. Later, when Joan was able to analyze her data about students' skills we could speculate about the reason for our concern.

Joan's assessment of the students' literacy skills revealed an interesting picture. Because of students' absences, she managed to track only 13 students' progress over the one-year period. On the one hand, the two students with extremes of attendance showed extremes of progress. One adolescent boy, whose attendance record was only 43% over the year, showed no progress in any of the key literacy outcomes. The Lanyons' young daughter's attendance was 97%, and she made excellent progress. On the other hand, the progress of the other students (whose average attendance was about 80%) was not closely correlated to their attendance. Instead, the results revealed that students' learning of English depended on the kinds of skills to be learned.

In the early years, to grade 2, students took longer than average to learn the conventions of text (the meaning of terms, such as word, sentence, letter, for example, and word spacing, letter-sound relationships, etc.). In the middle years, grades 4 to 8, students were slow to master the use of text and contextual understanding. We had seen examples where their literal interpretation of text impeded learning in other subjects, such as science and health. In the high school years, from grade 8 on, students were not staying at school long enough to move into the critical analysis of texts in relation to different social contexts. This meant that most students were not exposed to any extended knowledge of texts in relation to wider social issues and personal experiences. The "secrets," the hidden meanings, subtleties, and biases of the English language in texts, remained relatively unexplored by the Indigenous students at school.

The slowing-down process of learning to read English began early for students with English as their second language who mainly came from a non-print environment in the home. Time is the essence for learning, but trying to enable students to catch up required a carefully planned language program that was taught by highly skilled teachers. Catching up really meant that the students had to learn English at a faster rate, because nearly all of their learning of English had to be done at school. And this is a challenging feat for any student in any school.

There is no sense of blame here. The caregivers in the community supported the students' attendance at school, they helped with homework where they could, and they were pleased with what the school was doing. But, as in any community where English is not used out of school, and there are no texts in the home, it is difficult for students to learn everything about English they need to know to function as competent English users at school. What these results did support was the approach taken by the teachers to incorporate the use of language at every opportunity in their school curriculum.

Focus Questions

When English is the official language of a country, it is important that all students master it sufficiently to deal with everyday communication, including reading, writing, and speaking. The Indigenous students at Seaview Community School learned English as a second language, much the same as immigrants from a non-English speaking background. When students are immigrants, they also may not speak English at home and this may slow their language development.

1. Make a list of the advantages and disadvantages of the integrated approach taken by the teachers at Seaview. Discuss these with a colleague. How might some of the disadvantages be minimized?
2. In what ways did the teachers at Seaview assist students in reading, writing, and speaking English? Are there ways you can think of to increase students' use of English outside of school, when English is not spoken at home?
3. How can an integrated approach that focuses on language be detrimental to or supportive of learning in other subjects, such as science, mathematics, engineering and technology?

Suggestions for Further Reading

Fogelberg, E., Skalinder, C., Satz, P., Hiller, B., & Bernstein, L. (2008). *Integrating literacy and math: Strategies for K-6 teachers*. New York: The Guilford Press.

In this book, the authors focus on incorporating mathematics content into language and literacy. They argue that bringing reading, writing, and talking into mathematics lessons promotes the development of conceptual knowledge and problem-solving as well as computational skills. The authors explain what is meant by integrated instruction in these areas and show teachers how this can be done.

McKee, J. A. & Ogle, D. (2005). *Integrating instruction: Literacy and science*. New York: The Guilford Press.

This book shows how to use students' interest in science as a means of engaging them in motivating activities that develop literacy skills. It has a classroom practitioner focus with a range of activities that integrate science and other STEM subjects with literacy.

6

FOCUS ON REINFORCEMENT

Exploring Electricity and Energy Use at Beachville High School

Sheryl MacMath

Introduction

Curriculum integration in high schools presents unique challenges in comparison with earlier grades. Rather than having students in a single classroom with one teacher (elementary, ages 5 to 12 years), or continuous cohorts of students shared amongst a few teachers (middle school, ages 12 to 14 years), high schools have a mixture of students with separate teachers for each subject area. As a result, integrating multiple subject areas raises numerous scheduling difficulties. Not only are students typically taught by a different teacher for each subject area, but also students are rarely taught together in specific cohorts. In most high schools, students have varied class schedules with subject areas being taught at different times throughout the day. This results in scheduling challenges for high school teachers wanting to integrate across subjects such as science, mathematics, and social studies.

Given these significant challenges, one must wonder why high school teachers would attempt to implement an integrated unit. Some teachers who have implemented integrated units report that students are more motivated, that they feel more successful at learning, and that there are fewer discipline problems (Flowers, Mertens, & Mulhall, 2003). Here, we report on two teachers, Ms Wade and Mr Norris, at Beachville High School, who decided to implement an integrated unit on electricity and energy use. These teachers invited our research team into their classrooms. Interviews with both teachers occurred before, during, and after completion of the unit. In addition, we interviewed and observed eight pre-selected students over a period of five weeks: 19 visits with a total of 25 hours of observation time. Additional notes were kept of the classroom environment, teacher instructions, and levels of student involvement.

Beachville High School

Beachville has just over 500 students in grades 9 to 12 (approximately 15 to 19 years of age) in a large school district. In an effort to meet the aims of the state curriculum, Beachville offers a variety of subject areas for both academic- and applied-tracked students. Academic-tracked courses are geared toward students interested in continuing their education in a university degree program. In contrast, those courses that are tracked as applied are geared toward students interested in attending a college, trade school, or workplace preparation course. According to the state curriculum, the applied courses "focus on the essential concepts of the discipline, but develop students' knowledge and skills by emphasizing practical, concrete applications of these concepts and incorporating theoretical applications as appropriate." With the assistance of parents and guidance counselors, students choose to enter either the academic or applied tracks.

Ms Wade and Mr Norris taught grade 9 applied science and geography, respectively. In these classes, the majority of students were classified as "at-risk" students (the "at-risk" designation refers to students being at risk of not completing their high school diploma requirements). Aware of the need to provide at-risk students with instruction that is meaningful, personally relevant, and utilizing continual reinforcement (Johnson, 1998), Ms Wade and Mr Norris decided to coordinate their teaching of science and geography around the theme of energy use. The opportunity to try such an approach arose when the two teachers were scheduled on alternate days with almost the same students. Ms Wade and Mr Norris began by reviewing their curricula to identify commonalities. They noted that both curricula contained objectives related to electricity and energy use (see Table 6.1 for a list of objectives). Reviewing their schedules, both teachers decided to teach their electricity and energy use units starting mid-April, to conclude by the end of May. As a result, students would work with Mr Norris in his classroom on one day and these same students would then have Ms Wade in her classroom the next day. This approach represents a type of integration that may be described as synchronous: Ms Wade and Mr Norris were not teaching together, but, instead, were matching their lessons so that they would complement each other.

Prior to the start of the unit, Ms Wade and Mr Norris met formally twice—once over lunch and once for an afternoon (with substitute teachers provided by the school)—and exchanged several emails regarding their plans and expected timelines. During the course of the unit, while discussions with the research team did provide some indication concerning how each class was progressing (as the research team attended both classes for the unit), Ms Wade and Mr Norris were unable to meet formally. Following the conclusion of the unit, we had two lunchtime meetings that provided both teachers with the opportunity to discuss and reflect on the unit. Finding collaborative planning time was a considerable problem. However, even given these planning restraints, both teachers agreed that this was the best electricity and energy use unit they had taught, either independently or

TABLE 6.1 Objectives for the Integrated Unit on Electricity and Energy Use

Subject	Curriculum Objectives
Science	• Demonstrate an understanding of the principles of current electricity; • identify an authentic practical challenge or problem related to the use of electricity (e.g., to design household wiring, to increase the efficiency of electrical usage in the school); • design and build electrical circuits that perform a specific function; • describe and explain household wiring and its typical components (e.g., parallel circuits with switches, fuses, circuit breakers, outlets); • develop a solution to a practical problem related to the use of electricity in the home, school, or community; • compare electrical energy production technologies, including risks and benefits (e.g., explain the advantages and disadvantages of using hydro, photovoltaic, wind, and tidal generators to produce electrical energy); and • explain how some common household electrical appliances operate (e.g., electric kettle, electric baseboard heater, electric light bulb).
Geography	• Identify the locations and determine the relative importance of the country's major energy sources; • use selected criteria (e.g., costs, capacity, availability, sustainability, application, local attitudes) to evaluate alternative energy sources (e.g., solar, wind, tidal, hydrogen fuel cell) and conservation strategies; • describe the collective and individual/personal methods used in the community to reduce waste and conserve energy and water; and • explain the relationship between stewardship, sustainability, and change in consumption of energy (e.g., use of conventional versus alternative sources).

collaboratively. In the sections below we provide an overview of the unit, and its activities are described in Table 6.2.

Electricity and Energy Use

For this unit, students alternated between science and geography, with all science classes being taught by Ms Wade in the lab and all geography lessons being taught by Mr Norris in his classroom. While each teacher was responsible for teaching only their content area, they also referred to activities and content in the other class. For example, when teaching about conventional and alternative energy sources in geography, Mr Norris mentioned how electricity was moved from one place to another—a topic covered in the science class with Ms Wade. While the unit itself took five weeks to complete, the number of science lessons out-numbered the geography lessons three to one (see Table 6.2), due largely to the greater number of objectives for science. However, interestingly, it was the content covered in the geography class that kept reappearing in science. While Ms Wade's and Mr Norris's planning may have focused on providing their content in parallel, their actual sequence of activities wound up working progressively with the culminating

TABLE 6.2 Lesson Activities in the Unit on Electricity and Energy Use

Lesson	Subject	Activity Description
1	Geo	Working from a series of overheads, each student with his or her own mini booklet created by Mr Norris, students reviewed the different types of conventional energy sources. An introduction to turbines occurred with reference to science class. A country map was used to illustrate where these conventional sources are used.
2	Sc	Using a series of YouTube videos, students were introduced to how electricity gets from the generator to students' homes (e.g., transmission lines, distribution centers, etc.). Students then used digital cameras and walked around the outside of the school taking pictures of different electricity-related items (e.g., power lines, transformer, meter, etc.).
3	Sc	In groups, using pictures taken from the previous class as well as pictures Ms Wade had printed off the web, students created "electricity strips" showing the progression from power generator to the school. Students then presented their strips in class.
4	Geo	Working from a series of overheads, each student with his or her own mini booklet created by Mr Norris, students reviewed the different types of alternative energy sources. A map of the country was used to illustrate where these alternative sources are used.
5	Sc	Using wires, small light bulbs, and a battery, students worked in pairs to figure out how many different ways they could get the light bulb to work. Students were introduced to the terms power source, load, conductors, and circuit. Students completed a written lab report on what they learned.
6	Geo	After a brief review of conventional and alternative energy sources, students worked independently. Using an atlas, students worked on an individual map to identify all the different energy sources. In addition, students created their own map symbols to represent the different energy sources.
7	Sc	Using a demonstration of water through funnels, Ms Wade introduced students to the concepts of current, voltage, and resistance. Using the textbook, students then worked through a mini booklet created by Ms Wade on these concepts. When students experienced difficulty understanding resistance, Ms Wade played a game of follow-the-leader with and without chairs in the way so that students could physically simulate the concept of resistance.
8	Sc	Using a video, worksheets to complete, and a class discussion, Ms Wade overviewed electrical safety.
9	Geo	Using their symbols and individual maps from the previous lesson, students created a large wall-sized map of their country with the sources of energy appropriately labeled throughout. Then, after a brief review of the unit, students completed a summative quiz on conventional and alternative energy sources.
10	Sc	Working in small groups, students set up their own series circuit using two batteries, two lights, and several wires. Using voltmeters and ammeters, students compared measurements at different points in their circuits.

TABLE 6.2 (continued)

Lesson	Subject	Activity Description
11	Sc	Ms Wade demonstrated the creation of series and parallel circuits and related the use of these circuits to different rooms in a house. Additional worksheets were completed using the textbook (e.g., switches, fuses, etc.).
12	Sc	Using a series of PowerPoint slides, worksheets, and whole class discussions, students related the concepts of power, wattage, current, and voltage. Using a sample electricity bill, students were shown how these concepts related to the "cost" of electricity used in the home.
13	Sc	Using small group and whole class discussions, students brainstormed ways of conserving energy at home.
14/15	Sc	Working in small groups, using materials from around the science lab, students built 3D examples of how electricity moved from a power source to their homes. Groups were assigned a specific location and had to decide (and justify) why they chose a particular power source. Their justifications needed to relate both to availability as well as pros and cons of their energy source. In building their home they needed to include at least 5 ways they could conserve energy (e.g., hot water on demand, LED lights, etc.).
16	Sc	Unit review and test.

Note: Geo = Geography; Sc = Science

project linking back to the beginning of the unit (see Figure 6.1). As students worked through each set of science activities, they began recalling things they had learned in their geography class. Throughout the unit, three key examples of knowledge from geography class being used in science emerged. The first example occurred during Lesson 2. When Ms Wade showed video clips of different power generators, she asked students to identify the sources. Students quickly recalled the previous geography class, identifying them as nuclear power plants and "where they burn the coal" (thermal generators). Ms Wade was surprised that students knew what the pictures were of; however, students explained that they had seen them on the worksheets they had completed for Mr Norris.

The second major example of knowledge from geography making its way into science occurred during Lesson 11. During a whole-class demonstration, Ms Wade created a circuit using three light bulbs, wires, and batteries. Ms Wade noted that two light bulbs in the circuit were not as bright as the other one. She asked students to hypothesize why this was the case. One of the students quickly suggested that this was due to the fact that the duller light bulbs were further from the power source. When asked to explain why this would matter, several students started discussing how energy "leaks" when it has to travel a distance. When asked where they had learned this, students assumed that they had learned it in science class. When Ms Wade pointed out that this had not been covered in her class, it took students a moment to realize that this had been part of the work they had done with Mr Norris when discussing the location of energy sources in their country

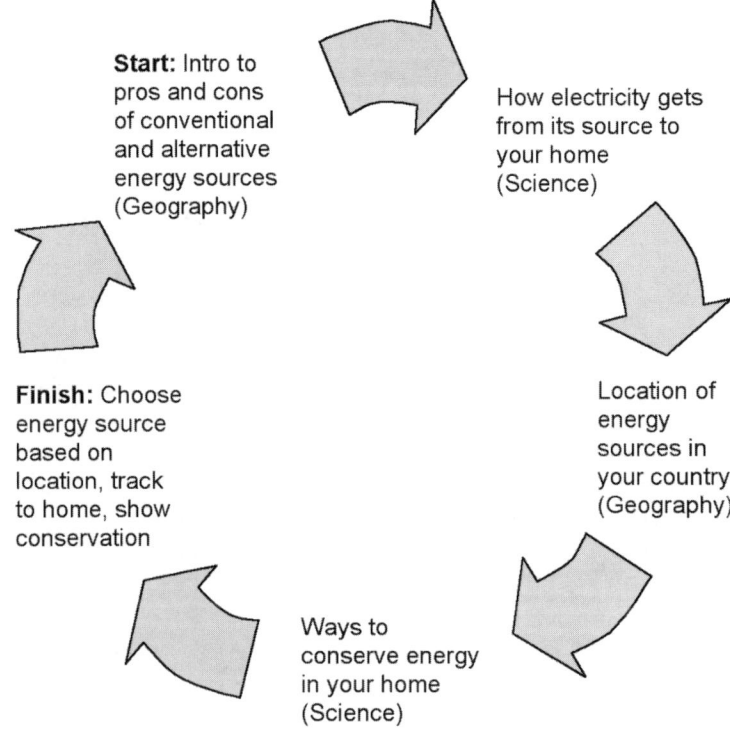

FIGURE 6.1 Sequencing in an Integrated Unit about Electricity and Energy Use

(Lesson 6). This example is particularly interesting in that it demonstrates how students were acquiring and using knowledge in a variety of settings without being able to recall exactly when or where it had been learned.

The final key example emerged in Lesson 13 when Ms Wade had students brainstorming different ways of conserving energy in the home. When one of the groups shared its list of possibilities it included getting their energy from "Bullfrog" (a power company that uses and markets energy from green sources, although at premium cost). During their geography class in Lesson 6, students had been introduced to an alternative energy source available to homes. Families and businesses could choose to get their electricity from "Bullfrog," considered a green (carbon-free) energy source using wind power and low-impact hydro as opposed to burning coal or nuclear energy. When Ms Wade enquired about including this suggestion, students discussed the value of using a green energy source as opposed to producing "lots of nuclear waste." Students were using what they had learned in geography, deciding that this was appropriate to their science discussion regarding energy conservation; "Bullfrog" represented a choice that they could make to reduce the impact of their electricity use on the environment.

In contrast to these unplanned examples of geography emerging in science class, Ms Wade consciously planned her final project to reinforce content covered in geography. The final project, given to students in groups of four or five, detailed the following:

> A new town has been created and your group of electrical engineers is in charge of getting electricity from a source to each home in the new town. Part of your job description includes using renewable or green energy where possible, and making sure the homes are energy efficient and are able to use electricity wisely. 1) Show how electricity will travel from a source to the town. 2) Show how electricity will be distributed to the houses. 3) Show how electricity will go into the houses and how it will be measured. 4) Design an energy-efficient house, listing five ways that the house will save electricity. 5) Create a name for your energy-conscious town. Present your concept to the class justifying your choice of energy production and explaining your model [handout from Ms Wade, Lesson 14].

Groups were assigned one of four locations chosen to give a variety of conventional and alternative energy available at these locations. By requiring students to work from a specific location and justify their choices, students needed to be able to demonstrate knowledge of the types of energy sources available throughout their country, as well as the pros and cons of different choices. None of this information had been covered in the science class. Instead, it represented an opportunity for students to use what they had learned in geography and connect it to what they were learning in science. Both teachers, while they could see the opportunities afforded by this approach, were interested in how best to assess students' understandings.

Success on the Final Project: What Does This Demonstrate?

Working with their students, self-efficacy was a major concern for both Ms Wade and Mr Norris. Students designated at risk of not making their diploma requirements typically demonstrate a lower self-esteem with regard to their academic work (Johnson, 1998). Aware of this, both teachers were interested in finding ways to help students feel good about their work; this was the primary reason these teachers decided to teach an integrated unit. However, they did not stop there. On the advice of another teacher at the school, both Mr Norris and Ms Wade decided to alter their testing practices. Rather than following a typical routine of reviewing content followed by a quiz on the next class day, both teachers decided to provide the review and the quiz on the same day (see Table 6.2). According to Mr Norris, "it's not like they're going home to study anyway." For both quizzes, the majority of students received scores higher than they normally attained. This resulted in a conundrum for both teachers: did students perform better because the content was integrated or because they had the quiz the same day as the review?

In their geography class, students completed their quiz during Lesson 9. The final project, where students demonstrated their knowledge of energy sources and the pros and cons associated with those sources, occurred during Lessons 14 and 15; this represented a time spread of two weeks. During their presentations students correctly identified the different types of conventional and alternative sources for their chosen location, and provided numerous key points regarding why they chose the source they did. Students demonstrated that they had retained the knowledge from their geography class well after the completion of the quiz. Mr Norris felt much more confident that the quiz scores were representative of student knowledge, more so than the timing of the quiz.

In contrast, Ms Wade's science quiz occurred after students had completed their final project. Consequently, her concerns were different from those of Mr Norris. During their presentations, students demonstrated a strong knowledge base regarding how electricity travels from a source to our homes, as well as ways of conserving energy, but these presentations were a very different format than a paper–pencil quiz. During their presentations students were able to respond verbally to questions, providing additional clarifications. They used visual examples from their 3D models as reference points in their presentations. Would students still do as well on a written quiz? Students out-performed expectations on the quiz (two students received 100%). However, at this point, Ms Wade was still left with the concern that student success was due to the timing of the quiz rather than knowledge acquisition. Would students have done as well if they had taken the quiz a few days later? Fortunately, our research team was able to answer this question.

Over the next several days, our research team conducted individual interviews with students. While asking them for their perceptions of the unit and each activity, we also tested students on the objectives outlined in Table 6.1. While these questions were answered verbally, students provided answers without their 3D models present and the questions were formatted differently than either quiz (both geography and science objectives were tested). All students performed at a level consistent with their scores on their respective quizzes. This illustrated that students were able to apply their knowledge to new situations even a few days after their class review. Given these results, both Mr Norris and Ms Wade decided that they would continue to give their unit reviews and quizzes on the same day with their applied classes.

Reflecting on the Unit

After the unit had finished, the research team interviewed both teachers. While Ms Wade and Mr Norris agreed that they needed more common planning time, especially during the course of the unit, both were pleased with their students' performance. They felt that the unit had flowed well, with each activity building on previous work. In comparison with previous work done by these students earlier in the year, both teachers felt that students had gained more knowledge and retained it for a longer period of time. Both teachers attributed this to the integrated unit. When asked why they felt that the unit had been beneficial to students, both teachers agreed that it was due to the reinforcement of content in two different classes. Students were introduced to concepts in one class that were used the next day in a different class. This is consistent with Nuthall's (1999) study of integrated learning, which revealed that students needed to have three or four experiences with a concept with no more than two days in-between those experiences in order to actually learn that concept. Given their schedule of alternating class days, Ms Wade and Mr Norris were able to provide those conditions for their students. According to both teachers, "this made all the difference."

Interviews and surveys with the students after completion of the unit echoed the teachers' perceptions. Students agreed that it was very helpful having content covered in both classes. They commented on how "things [Mr Norris] said matched what we did in science" and how it was "easier in [Ms Wade's] class because we already knew about it from [Mr Norris]." In addition to finding the reinforcement helpful, questions on the surveys revealed that students felt they were "more successful in this unit than other units." Students' own perceptions of their academic abilities were high. They agreed that they "worked hard on this unit" and that they "did well on this unit." Students' feelings of self-efficacy were positive—a key goal for both teachers.

We conclude that the kind of integration we saw in this unit on electricity and energy helped reinforce important concepts. This reinforcement assisted students to experience success, and, importantly, feel good about their success. Ms Wade commented, and Mr Norris agreed, that while "this would be good for our academic classes, it's especially important for our applied students." For these at-risk students, the reinforcement of ideas on a regular basis in a variety of settings made a difference; synchronous integration was a valuable tool, which helped to increase student success and academic self-efficacy.

Focus Questions

1. A lack of common planning time was a real problem for both teachers. What other strategies could have been implemented by these two teachers to keep each other informed of their progress?
2. In some instances, when teachers team-teach a unit, one subject or discipline is dominant. That was the case here, with science being more dominant that geography. Can you identify any concerns that may arise by having one subject dominant over others? How, as teachers, can we address these concerns?

3. In this case story, when information was repeated and reinforced, the setting was very different: different classrooms, different teachers, different textbooks, etc. What are the advantages and disadvantages of having this unit taught by one teacher in a single classroom?

Suggestions for Further Reading

Applebee, A. N., Adler, M., & Flihan, S. (2007). 'Interdisciplinary curricula in middle and high school classrooms: Case studies of approaches to curriculum and instruction'. *American Educational Research Journal*, 44(4), 1002–1039.

This research study of classrooms in two US states draws comparisons between different types of curriculum integration, and presents some useful models for conceptualizing integration.

Nuthall, G. (1999). 'The way students learn: Acquiring knowledge from an integrated science and social studies unit'. *The Elementary School Journal*, 99(4), 303–341.

This is an accessible article presenting some key points about student learning in an integrated project involving science and social studies.

References

Flowers, N., Mertens, S. B., & Mulhall, P. F. (2003). 'Lessons learned from more than a decade of middle grades research'. *Middle School Journal*, 35(2), 55–59.

Johnson, G. M. (1998). 'Principles of instruction for at-risk learners'. *Preventing School Failure*, 42(4), 167–174.

Nuthall, G. (1999). 'The way students learn: Acquiring knowledge from an integrated science and social studies unit'. *The Elementary School Journal*, 99(4), 303–341.

7

FOCUS ON FOCUS

Making and Marketing a Toy at Rinkview Public School

John Wallace

Introduction

The case described in this chapter is set in Rinkview Public School, a senior elementary school (grades 6 to 8) in a large suburban school district. The focus of the case is a grade 8 pod consisting of two teachers, Ms Charlie and Ms Jane, and their 50 students who were engaged in a six-week integrated unit of work called "Making and Marketing a Toy." The unit encompassed a broad range of content and skills, drawing from state curriculum expectations in science, mathematics, design technology, and social studies, as well as a number of interdisciplinary goals directly related to the task. In conducting this study we were interested in the integrated nature of the unit and the relationship between disciplinary, interdisciplinary, and other knowledges. We wanted to know how the two teachers managed these different components of the curriculum, and, importantly, what the students learned in the process. We made regular observations and interviews over the six weeks, focusing our attention on the two teachers and 12 of the students who were working in three groups of four.

The Unit Activities

Ms Charlie and Ms Jane based their planning for this unit on an integrated unit plan provided by a teacher in another school. The unit, originally titled "Create Your Own Business," involved the integration of several subjects, including language arts, fine arts (posters, displays, radio and TV announcements, advertisements), social studies (marketing, global economy, consumerism), and mathematics (money management, use of charts and graphs, central tendencies). Ms Charlie's particular interest was in mathematics, science, and technology, while Ms Jane specialized in social studies and language arts. The two teachers, working in collaboration with the six other grade 8 teachers at Rinkview, decided to modify the unit to include science. As part of the unit, students were expected to design their own moving toy and to explain that movement in terms of the science of simple machines. All grade 8 classes at the school were involved with the unit with a culminating toy fair containing over 40 different toy designs. The grade 6 and 7 students in the school attended the toy fair and, using a paper-and-pencil debit system, bought the toys they preferred; the toy design making the most money would be declared the "winner."

While this unit touched on a number of subjects, our focus was on the integration of science, mathematics, and technology. Prior to the unit, the two teachers specified two "integration" objectives: (I1) identifies connections between school activities and real life; and (I2) identifies connections between subjects. There were also several disciplinary goals taken from the state curriculum guidelines. These goals included three science-related objectives: (S1) demonstrates understanding of the factors that contribute to the efficient operation of mechanisms and power; (S2) uses appropriate terminology to communicate ideas; and (S3) communicates procedures and results of investigations for specific audiences. There were two mathematics-related objectives: (M1) uses information from charts and graphs to make decisions; and (M2) knows and uses measures of central tendency to make decisions. There were two technology-related objectives: (T1) recognizes the needs of consumers when designing; and (T2) uses and values the design process.

The unit was broadly organized around seven classroom activities, designed by the teachers to scaffold student learning toward the final goal of building and marketing a toy, and also to meet the various unit objectives outlined above. The seven activities were lever experiments, sentence strips, consumer surveys, survey analysis, schematic designs, toy construction, and toy fair.

Activity 1: Lever Experiments

For this first activity, students were asked to use meter sticks, books, and desks to create and test class one, class two, and class three levers. Students experimented with different distances between the fulcrum and the load to see if the distance made any difference to the effort required to move the load. The intention here was to help provide students with some of the science background knowledge behind the workings of their future toy.

Ms Charlie explained how the students were to position their meter sticks (lever), books (load), and desks (edge being the fulcrum). Students seemed uncertain about where to place the load and whether they were to alter the position of the load or the fulcrum to determine which positions were the most efficient. Although Ms Charlie stopped the class several times during the lesson to clarify her intentions, some students remained confused, resulting in different conclusions by each group. In Ms Charlie's final interview, she acknowledged this confusion: "translating those written instructions into an action was the challenge for them."

The lever experiments focused primarily on the first two science objectives (mechanical efficiency and using correct terminology). Both teachers circulated around the room asking each group key questions about their results. In reflecting on the lever experiments, both teachers said that the students really enjoyed the activity "because they got a chance to do it themselves, rather than [just] showing them or telling them" (Ms Charlie). When commenting on the confusion about the experiments, Ms Charlie said that "they learned a lot, because we had a lot of misunderstandings, even after the experiment was conducted ... and that misunderstanding led to clarification." This clarification occurred during the next activity when students were asked to organize their results into one or two sentences.

Activity 2: Sentence Strips

To consolidate students' learning from the previous activity, Ms Charlie instructed students to write one or two sentences to "sum up" the information they had gathered on the three classes of levers. Recognizing that students had drawn different conclusions from the experiments, Ms Charlie wanted to create an opportunity for students to compare results, challenge each other's conclusions, and come to a whole class consensus. "That's why I think that the community group discussion that we had them doing is so important, because it just clarifies" (Ms Jane).

Students sat in a large elongated circle that stretched the length of both classrooms. Ms Charlie began by asking students to "spontaneously jump up" (she called this a "popcorn" strategy) to share what they learnt about class one levers. However, after repeated attempts at encouraging students to share verbally and independently with the class (including an explanation of the popcorn metaphor), few students volunteered to speak. As a result, Ms Charlie asked a representative from each group to read out what they had learned from each activity.

After the groups shared their results, Ms Charlie reviewed the different experiments with the whole class. Two key learning outcomes were identified. First, students agreed that, for class one and two levers, the closer the load was to the fulcrum, the more efficient (the easier) it was to lift. Second, students also began identifying a variety of levers that they used in everyday life (e.g., stapler, bat, etc.). Many students commented on how this was a new idea for them; they had not realized that levers existed all around them.

Activity 3: Consumer Surveys

For this activity, Ms Charlie instructed students to design a survey to identify what consumers (in this case, the students in grades 6 and 7) wanted in a toy; if they were to be successful in selling their toy, they needed to build something that consumers would buy. A brief period of time was spent discussing the types of questions that might have been relevant to making their toy. Questions regarding price, color, function (e.g., rolls, waves, speaks, etc.), and type (e.g., car, action figure, etc.) were proposed. Students worked in their groups for 45 minutes to finalize their questions, and Ms Charlie instructed students to gather the data before school, at lunch, and after school.

When the groups reconvened a few days later to share what they had learned, there was much discussion about the appropriateness of the questions. For example, most groups surveyed students with the question, "How much would you be willing to spend?" on a toy. Responses ranged from $15 to $50. However, this question had been given without context, thus creating a problem. For example, while students might be willing to spend $60 on a video game, groups were not going to be making a video game. As a result, they were unable to use some of the data they had gathered.

> After they had completed their own surveys, some of them realized that their questions may not have been as appropriate to their needs. And they wanted to go back and re-do the survey. And, well, that's great, but we don't have the time, unfortunately (Ms Charlie).

During this whole-group discussion, Ms Charlie emphasized the importance of relating their survey questions to "toys they could actually make," and the first integration objective (relating school activities to real life) took on some significance. "It's not the first time they were doing the survey, but it was the first time they were critically thinking about the questions that they were asking" (Ms Charlie).

Activity 4: Survey Analysis

After providing an overview of their survey results to the class, Ms Charlie instructed groups to draw a graph of their results (using a graph of their choice) and determine the measures of central tendency for each of their questions. While students got to work quickly, they did not appear to understand how this activity related to the making of their toy; it was as if this was a separate task, unrelated to the toy unit.

There was mixed success when working on the graphs for each question. Group One needed additional support as they tried to turn response counts into percentages and attempted to represent those percentages on a pie graph. Group Two wanted to show responses to all questions on one chart. To do this, they decided to only display the response that had received the most votes. However, this visual representation did not reflect the variety of responses their questions had elicited. Group Three created four different charts, one for each question, and calculated the mean, mode, and median for each question.

After 50 minutes of group work, Ms Charlie had one of the groups present their chart using a document imager to project students' work on to the wall. This group was having difficulty determining their measures of central tendency. The group members asked, "When your question asks what color would you prefer and your mean is 25, what does that tell you? Does that mean anything?" Other students provided responses to this question; the discussion revolved around having 100 responses divided by 4 (blue, red, yellow, black) always gives 25, regardless of the question. The teacher then explained the difference between mean, mode, and median. She pointed out that for categorical data, involving no measure of scale, the mean and the median are irrelevant—only the mode can be calculated. The mode, referring to the response that had been chosen the most, gave a categorical response (e.g., black is the most preferred color).

Group Two presented their consolidated chart to the class, demonstrating how its chart reflected the mode for each question. Discussion then arose regarding the loss of information in the consolidated chart and the advantages of using a bundled chart that stacked responses for each question, enabling all responses to be represented on one chart. There was a further discussion about medians and how questions that involved price selection could be ordered from least to greatest (e.g., $5, $10, $15, $20), enabling a median to be calculated.

In the final student interviews, students referred to these discussions, using the terms mean, mode, median, and categorical data. While sometimes confused about the name for each measure of central tendency, the majority of students successfully calculated each measure in their final interview. Several students said that they found the calculation of central tendency helpful in making their toy decisions, and that the visuals were useful guides.

Activity 5: Schematic Designs

For this activity, each group decided on the kind of toy they were going to make and drew a schematic of the toy. In the drawing, students were expected to demonstrate the scientific principles that enabled their toy to move. These criteria were based on a rubric that Ms Jane first shared with the class. After agreeing on the design of the toy, they generated a materials list and expense report. Ms Charlie reminded students that they should be able to back up their design with information from their surveys. She used a metaphor of varying brands of MP3 players to introduce the idea of a "made-from-scratch" toy as opposed to a brand new idea for a toy. She also introduced the idea of a prototype.

During the initial brainstorming session, Groups Two and Three came up with several rough schematics. After observing students' initial drawings, Ms Charlie reminded them again that they could not modify an existing toy, but had to make one from scratch. She said it had to be their own idea, suggesting that toy websites could be a good resource. Some students asked if they had to build everything by themselves. Ms Charlie then responded that, "If you need to add a piston, do you make it from scratch? No … get a small part. … " Ms Charlie then used the notion of a "Frankenstein-ed toy" assembled by bringing spare parts together. Students then asked if they would have to build a motor from scratch and Ms Charlie responded, "No, but you will have to be able to explain the scientific principles involved with your motor."

This discussion wound up being critical to the success of Group Two. The majority of this group's members interpreted this discussion to mean that they could use, or "Frankenstein," the undercarriage of a remote-controlled car and then build their own body, adding in various electrical circuits for the control of lights and music. At the end of the unit, the teachers penalized Group Two because they did not build their toy from scratch. In their final interview, Ms Jane and Ms Charlie revisited this issue:

MS CHARLIE: But at the very last minute, they took an existing remote-control car. Well, it seemed to me, at the very last minute, because I didn't get a chance to see their plans, unfortunately, and that, you know, I think is probably our fault as well, too, with just how busy we were. We just kind of … we touched base with them.

MS JANE: But even so, that just came in towards the end. Because I was under the impression that they were making something totally from scratch.

MS CHARLIE: So did I, because when I talked to them … they knew they had to Frankenstein their work. And they ended up going with a remote-control car, taking the body off, and creating their own body. And they added the speaker, and stuff like that.

RESEARCHER: And the lights.

MS CHARLIE: And the lights. But that wasn't the idea … to take an existing toy and product and modify it. It was to create it from scratch.

As illustrated by this discussion, even the teachers had different interpretations of the notion of a "Frankenstein-ed toy."

In their deliberations about toy design, students were encouraged to take into consideration such factors as mechanical advantage (S1), survey results (T1), aesthetic appeal (T1), expenses (I1), and materials (I1). Group One decided to make a car that could transform into an action figure. They thought they might use elastics to get the car to move. Group Two decided to make a hovering craft. One group member, Mason, explained that they planned to use a motor that "is attached to a gear which is attached to a fan." He also suggested attaching a "cheap walkie-talkie" to the craft. Group Three decided to make a Snake-a-saurus whose tail and eyes would move. Each group kept a record of progress in their journal.

Ms Charlie reflected on this activity by pointing out that "the groups that need [ed] a lot of scaffolding … wanted to rush right into creating the product." Ms Jane struck a similar note when she said that this activity was very difficult for this age group because "they don't want to sit down and actually plan something." She also observed that the students began to better appreciate the design process by the end of the project. This point was echoed in the final interviews with some of the students mentioning how the schematic activity helped them sort out ideas. Upon reflection, Ms Jane said that if she did this activity again, she would make some changes: "I would actually have them go home and try to bring in a blueprint of something. To actually show them what a blueprint is." In the end, all of the schematics would require modification during the construction phase.

Activity 6: Toy Construction

On construction day, the entire day was set aside for toy making. Ms Charlie provided each group with a task sheet to record hourly activity. She reminded students that, over the course of the day, they were to construct (and finish) their toy, complete an expense report, and build a display board. Students were expected to have all the necessary materials at hand and to build their toy based on their schematics.

Students worked in groups for this activity. After the lunch break, there was a whole class discussion about the expense report. Ms Charlie and Ms Jane emphasized that "nothing is free. I know you think it grows on trees, but it doesn't." Ms Jane told students that all the materials the students used should be included in their cost. The noise level for this activity was quite high and, from time to time, students walked around the classroom, observed other groups, traded materials, or engaged in off-task talk. Some groups also experienced many challenges with their schematics and materials.

Of the three groups observed, Group One had the most success on construction day. While members had originally planned to build a toy car, as a result of disagreements regarding type of building materials, concerns that other groups were also making a car, and that a car would be too difficult (only William felt confident in building and using a motor), they changed their minds prior to construction day. Instead, they decided to build a Buzz Fly, which involved attaching a Styrofoam ball to a long lever that could be used to propel the ball through the air. Group One's idea came from a toy similar to one that Alex already owned. Alex and Rihan worked on the toy and William and Samir worked on the display board. They completed the construction of their toy in the morning, along with test trials in the hallway, and in the afternoon completed their display board.

Group Two spent the morning of construction day working on building an ICow, a vehicle that rotated while moving in a straight line. While Group Two brought all the materials it figured was necessary, it experienced difficulties with the rotating wheels, deciding, after much work, that it needed an additional gear and battery. As a result, they were not able to build the base of their toy. By the end of construction day, Group Two had not made any headway on their toy, their display board was incomplete, and they had only two posters for distribution; it had not been a productive day. Group Two met over the weekend and, after much negotiation, decided to not continue making the ICow as they felt they did not have enough time to complete it. Instead, they took apart and used the undercarriage of a remote-controlled car, built a removable outside carriage out of popsicle sticks, and connected batteries and wires to both a lighting system and an MP3 player.

Group Three brought the majority of their supplies to build their Snake-o-saurus on construction day. However, they had assumed that they would be able to cut and mold chicken wire with regular scissors and glue. Within the first half hour, they realized that this was impossible. Travis suggested that they meet at his house on the weekend to use his dad's tools to cut and shape the chicken wire. For a brief period of time, the group attempted to get the wheels to move using a pulley; however, they were unsuccessful in getting their wheels to work. For the rest of the day, Group Three spent some time working on their display board and posters. However, the majority of the day was spent in off-task discussions with other groups and, as a result, very little was accomplished on construction day. Over the weekend, Group Three met at Travis's house as planned. They decided to papier mâché the Snake-o-saurus, which Travis finished over Monday and Tuesday. The rest of his group finished the display board at Caitlin's house.

Activity 7: Toy Fair

The toy fair took place in the school gymnasium, where over 40 toy stations were set up by grade 8 students. The purpose of the activity was to market the toy and make as much profit as possible. Grade 6 and 7 students were brought in to shop using a paper debit sheet. Each student could "spend" $50 on toys. Each toy station kept a separate account of how many toys were sold and how much money was made. Using the expense reports they had already created, the grade 8 students would then subtract their total expenses (including advertising) from their total profit to determine net gains; the group with the highest net profit would be declared the "winner" of the fair.

As the fair proceeded, it was clear that some students did not realize that there was a $50 limit placed on their consumers, thus affecting their potential profit margin. An additional problem was experienced by Group Two. When the teachers observed their remote-controlled car and quizzed them on its construction, they penalized the group for not having made their toy from scratch, forcing the group to change their price from $38 to $50. Livia from Group Two felt that "now it's too expensive and [we] don't think anyone is going to buy it." In their final interviews, students from Group Two were still very upset about this decision.

Based on general observations from toy fair day, most students were very engaged and enjoyed the "selling" process. In their final interviews, students commented on how much they enjoyed toy fair day. They saw it as "the most important stage of the toy-making project … the entire project was based on this one day" (Tonya). All students cooperated well in their groups for this activity, dividing up the labor—some students solicited potential consumers, some demonstrated the functions of the toy, while other group members kept the sale records.

According to the teachers, the toy fair day was focused on an integrated objective (connecting with real life). This aspect was also emphasized by students in their final interview. They saw it as a type of test, the opportunity to see what the "public" thought of their toy idea. In reflecting on this culminating activity, Ms Charlie also mentioned how the experience taught students about real life: "It's a real-world example of what's happening, right. We want to make money. How are we going to make money? We have to sell our product. Well, we have to create a product first that people are going to want to buy."

Discussion

The Rinkview case story reveals both the potential and the challenges of this kind of curriculum integration. The teachers are to be commended for their work in bringing the subjects together in such an integrated manner. In terms of student motivation, the students in this classroom were well engaged, as indicated by very high levels of on-task behavior. They expressed enjoyment, saw the relevance and importance of the work, appreciated the integrated aspects of the unit, and generally liked working in groups to solve problems. We also found evidence of students' capacity to transfer general problem-solving skills. There were also some good examples of students bridging subject knowledge to the task at hand, although many struggled to make the connections expected by the teachers.

Part of the challenge of this kind of teaching, however, is related to the issue of focus. According to Ross and Hogaboam-Gray (1998), focus is one of the arguments for integrated curriculum, "based on the belief that students are more likely to learn when their attention is focused on a few objectives or big ideas rather than diffused among many" (p. 1120). These authors referred to three aspects of this argument. First, that integration can potentially focus on a few big, shared ideas or topics, rather than a myriad of smaller and often unrelated discipline-based concepts. Second, because some subjects are supposedly complementary, tools from one subject can be used to enhance learning in another subject area. For example, science and mathematics may contribute to technology learning by providing students with some conceptual and process skills to solve technological problems (and vice versa). A final aspect of the focus argument concerns the potential for integration to highlight the quintessentials of each subject. It is suggested that, paradoxically, teaching in an integrated manner may actually assist students to comprehend and understand the traditions of individual subjects, including knowledge organization, key concepts, what counts as evidence, and truth warrants.

In examining the Rinkview case, we found that the issue of focus is more complex than imagined within the literature. Rather than emphasizing clarity, purpose, complementarity, and quintessentials, the toy unit was characterized by a multitude of different goals and expectations. The teachers, for example, hoped that the units would serve some of the purposes set down by the state curricula (in terms of science, mathematics, technology, etc.), and help the students achieve some cross-curricular goals (such as applying knowledge to "real life"), as well as complete or build the final project (the toy) and participate in a competitive marketing exercise. At different times during the units, the teachers emphasized different aspects of the unit. For example, after the students had conducted a peer survey on the marketability of their toy, the teachers taught a lesson on the mathematical principle of central tendency to help students organize their survey data. The teachers also taught formal lessons on the characteristics of levers in the hope that this knowledge would be applied in the construction of the toy. In building the toy, the students were asked to focus on several aspects, including design, aesthetics, workability, and marketability. Moreover, the technological goals of making a working toy were conflated with the economic goals of making a profit.

For the students, this diffuseness of goals led to some focus confusion. While there is strong evidence of high levels of student learning and engagement, at times the tasks seemed more fragmented than integrated. Sometimes, students were unsure whether to focus their energies on the final product or to spend time on the specific tasks set by the teachers along the way. Teachers were also dealing with multiple goals, citing state curriculum expectations, broad goals such as critical thinking, as well as practical task expectations. The Rinkview teachers expected the students to complete various tasks, including classifying levers, drawing graphs to report the results of the market survey, and detailing the trials of the performance of the toy. When it came to the final product, it was somewhat unclear whether the principal focus was on the design and build aspect or the marketing aspect of the toy. Moreover, some students were unsure about the features of the final product; did it have to be designed from the ground up, for example, or could it be an adaptation or modification of an existing toy—a "Frankenstein-ed" toy?

Part of the problem here, we suspect, lies in diffuseness of focus. The teachers (and hence the students) were somewhat conflicted about the goals of the units, and uncertain at times about whether to focus on final products or other related (mainly disciplinary) knowledge and skills. Students and teachers were also pulled back and forth between knowing and doing, disciplinary and inter/transdisciplinary, mathematics and science, technology and economics, compulsory and optional (in terms of mandated expectations), fun and serious, etc. While these tensions may be resolved in part by a clearer task definition, we suspect that focus uncertainty is part of the continuing landscape of curriculum integration.

Conclusion

We suggest that student learning success in this case is related to the issue of focus. Some degree of focus diffuseness is part of the unpredictable nature of integrated classrooms. The challenge is to avoid focus confusion, to be clear about the parameters and expectations of the task/s at hand while allowing plenty of space for student creativity. What this requires is a kind of meta-awareness in integrated teaching, learning, and assessment—including continuous, clear, and explicit attention to goals and purposes. At times, focus will also mean being explicit about some of the subsidiary goals of the unit (such as certain subject matter understandings). Learning is enhanced when goals are clearly stated and when assessment is aligned with those goals. The challenge as always is to identify the things that matter, make those things explicit, and to align teaching, learning, and assessment practices with the most important conceptual and procedural ideas and activities.

Acknowledgement

Sincere thanks to Sheryl MacMath and Xiaohong Chi for their invaluable assistance in collecting and analyzing data for this case story, and producing project reports.

Focus Questions

1. In your opinion, what are the "things that matter" in this integrated unit (that is, what should the teaching and learning be focused on, what are the big ideas, the little ideas, etc.)? How do you decide what matters and what doesn't matter?
2. What teaching strategies would you use to focus students' attention and make explicit the things that matter?
3. Assessment is one of the biggest challenges in a unit such as this. What are the most appropriate assessment tools for a unit such as this? How would you ensure that teaching, learning, and assessment are appropriately aligned?

Suggestions for Further Reading

Drake, S. M. (2007). *Creating standards-based integrated curriculum: Aligning curriculum content, assessment and instruction* (2nd ed.). Thousand Oaks, CA: Corwin Press.

This resource text offers a framework allowing for multidisciplinary, interdisciplinary, and transdisciplinary approaches to curriculum integration, provides sample models of integrated curriculum in action, gives some practical suggestions to simplify curriculum alignment and integration, and provides insights based on classroom experience to connect the framework to the real world.

Wallace, J., Venville, G., & Rennie, L. (2010). 'Integrated curriculum'. In D. Pendergast & N. Bahr (Eds.), *Teaching middle years: Rethinking curriculum, pedagogy and assessment* (2nd ed., pp. 188–204). Sydney, NSW: Allen & Unwin.

This chapter looks at the different conceptions of curriculum integration, reviews some of the types of integration based on classroom observations, looks at how the various knowledge interests may be pursued in integrated teaching and learning, and proposes a teaching planning framework based on the "Worldly" curriculum perspective.

Reference

Ross, J. A. & Hogaboam-Gray, A. (1998). 'Integrating mathematics, science, and technology: Effects on students'. *International Journal of Science Education*, 20(9), 1119–1135.

8

FOCUS ON TEACHER SUPPORT

Considering Access for the Disabled at Gosport Community School

Rachel Sheffield

Introduction

This chapter explores the barriers that hindered an early career science teacher as she implemented an integrated project. The case story is about a project called "Making a Difference—Community Access for People with Disabilities" that was implemented across several subjects in two grade 8 classes (students aged 12 to 13 years). We highlight the frustrations the teacher, Ms Potter, faced when struggling to collaborate with colleagues, negotiate with the school administration, and engage and motivate the students in her science classes.

A major goal for science education espoused in curriculum documents throughout the world is that students develop an understanding of how science and society are mutually dependent. Successive documents in the United States (American Association for the Advancement of Science, 1993; National Research Council, 1996, 2011), for example, emphasize the need for science, technology, and more recently engineering, and point out that these disciplines "are not only major intellectual enterprises but can improve people's lives in fundamental ways" (National Research Council, 2011, p. 1). The purpose of the project described in this chapter was to bridge the divide between science and the social sciences by enabling students to understand how the application of scientific and technological knowledge (such as simple machines, levers, ramps, friction, etc.) impacts on the lives of disabled people living in their immediate community. The data collected during the case story suggested that, while curriculum integration may be a potentially appropriate solution to the contextualization of science concepts within social issues (Lewis & Leach, 2006), this project may not have made a difference to the engagement and participation of the students in science. We noticed a number of factors that faced the science teacher which became barriers to the implementation of the integrated project.

Background to the Study

Gosport Community School has approximately 1600 students housed in a purpose-designed middle school (grades 8, 9, and 10) and a senior school (grades 11 and 12). The middle school and senior school run to different timetables, although all students share common teaching areas. The school campus has a large library that is not only available for the students but also serves the general public. The middle school has a number of learning teams each comprising six teachers, who teach content within their specialized subjects (e.g., mathematics, science, and English). A nominated team leader is responsible for coordinating each learning team. The teachers in each team share a small office that is in the center of a cluster of classrooms. Ms Potter's learning team included five other more experienced teachers and was led by the social studies teacher, Mr Birchwood, who was a very experienced teacher.

The grade 8 integrated project, "Making a Difference—Community Access for People with Disabilities," was initiated by a parent who, as a member of the school board, had a longstanding professional association with the school. The parent, Ms Mills, worked within the community with families of young people who had a range of disabilities. She received a small government grant to examine disabled people's abilities to access community facilities, including shopping centers, beach fronts, and parks. She approached the school, through Mr Birchwood, to help her with this project. As a consequence, students in grades 8 and 10 became involved in the project. The grade 8 students focused on community access for people with disabilities and the older students designed a calendar and website to assist carers to locate access sites for people with disabilities.

The young science teacher, Ms Potter, was in her second year of full-time teaching, having completed one year of science teaching in a rural school. This was her first year at Gosport Community School. Ms Potter was assigned to the middle school, teaching two grade 8 classes, and four grade 9 classes. She was not familiar with all the science topics taught in the middle school, some of which she was teaching for the first time. She taught most of her middle school classes, including both grade 8 classes, away from the learning team's classrooms in a science laboratory in the main science building.

Case Story: "Making a Difference"

Over the course of the term, in my role as researcher, I observed the majority of Ms Potter's grade 8 classes. I also attended two learning team meetings, the disability workshops, and activities that included several grade 8 classes, led by Mr Birchwood and Mr Smith, the physical education teacher. I was also present for the end-of-term excursion to visit the "Wheel Cats" disabled basketball team. Table 8.1 provides a list of the subject content and the teaching and learning activities in which the students in one of Ms Potter's grade 8 classes participated during the project.

TABLE 8.1 The Content and Teaching/Learning Activities Used in Lessons Observed in Ms. Potter's Class over the 11-Week Project

Week	Lesson	Subjects	Content	Teaching/learning activities
1	Lesson 1 (½ day workshop)	Science and social studies	Disability Education Sports Program The senses, injuries, and wheelchairs	Guest speaker from Disability Education Sports Program Group work of interactive activities. Worksheet and reflective journal
	Lesson 2	Science and social studies	Round robin of questions relating to disability, for example, How do wheelchairs assist people? What problems do people in wheelchairs face?	Group discussion of question and share comments with class
2	Lesson 3 (full day workshop)	Science and social studies	Thoughts and feelings about disabilities	Guest speakers, trust activities, and exploring the school
	Lesson 4	Science	Forces	Textbook used to complete poster on forces
3	Lesson 5	Science	Forces—students had misbehaved in the lesson before so repeated the lesson	Textbook used to complete poster on forces
	Lesson 6 (½ day workshop)	Science, physical education, and social studies	Disabled sports	Group activities on oval Guest speaker from Disability Education Sports Program
	Lesson 7	Science	Forces	Textbook used to complete poster on forces
4	Lesson 8	Science	Machines and forces	Teacher dictation of definitions using the textbook
	Lesson 9	Science	Force and speed in simple machines	Students complete worksheet using website
	Lesson 10	Science	Tests on types of simple machines	Formative assessment worksheet on simple machines

TABLE 8.1 (continued)

Week	Lesson	Subjects	Content	Teaching/learning activities
5	Lesson 11	Science	Robot kits	Students placed in groups and collect and record details of Lego robot kits
	Lesson 12	Science	Robot kits	Lego kits to continue to try to build robots
6	Lesson 13	Science	Robot vs. machines	Lego kits used to continue to try to build robots
	Lesson 14	Science	Links with disability unit	Textbooks used to answer questions from worksheet
7	Lesson 15	Science	Uses for robots	Textbooks Think/Pair/Share
	Lesson 16	Science	Substitute teacher provided to allow collaboration with social studies teacher	Students use books to answer questions from worksheet
8	Lesson 17	Science	How robots are used	Mini-reports prepared using data collected from web
	Lesson 18	Science	Create model using Lego	Create and build model following instructions
	Lesson 19	Science	Create model using Lego Program Lego model	Program model using computer
9	Lesson 20	Science	Friction robot investigation using Lego model	Investigation—review of findings Investigation scaffold, teacher reviews, students copy down answers
	Lesson 21	Science	Friction robot investigation	Investigation using scaffold (continued)
	Lesson 22	Science	Science investigations on friction	Investigation using scaffold (continued)
10	Lessons 23	Science	Science investigation on friction write up	Computers used to write up investigations
	Lessons 24	Science	Science investigation on friction write up	Computers used to write up investigations
11	Lesson 25 (full day excursion)	Science and physical education	Wheelchair basketball	Excursion: Students learnt how to play wheelchair basketball with the Wheel Cats Basketball Team.

Her second class participated in a very similar program of work. The science taught through the project was based on a topic in grade 8 previously called simple machines. Students were able to construct, program, and then use Lego robots to traverse different surfaces and plot the time and distance coordinates as scientific investigations. The investigations could be used to connect with disabled people's problems with wheelchairs, in particular, the problems of traveling on various surfaces and moving up ramps. The robot investigations also could be used to demonstrate the use of levers to help people with disabilities.

Over the course of the term, data were collected through informal conversations with students and Ms Potter, a formal interview with her, examination of artifacts, field observations of classroom lessons and joint activities with social studies (see Table 8.1), and a survey of the grade 8 students. The findings revealed a number of factors that existed within the classroom setting that impeded the teacher's ability to deliver the integrated project. These factors are discussed under the headings: collegial collaboration, school administration, teacher inexperience, teacher isolation, and disaffected students.

Collegial Collaboration

Ms Potter's colleagues lacked interest in adopting and participating in this project. The project was initially designed at a team meeting to be a collaborative, integrated project, and to incorporate all subject areas under the theme of community access for the disabled. My observations revealed, however, that the project was not implemented by all members of the team. The only teachers who made a sustained contribution were the science teacher, Ms Potter, the learning team leader who was also the social studies teacher, Mr Birchwood, and to a limited degree, the physical education teacher, Mr Smith. Table 8.1 provides more detailed information about the involvement of the different subjects in the project.

The remaining three teachers in the learning team simply continued to teach their previously established programs of work. Ms Potter expressed her understanding of the other teachers' perspectives: "don't get me wrong, it takes energy [to include the theme] and the others expect to be told what to do." These other teachers also maintained an independent approach to teaching their subject matter. In the interactions during the learning team meetings, the majority of the discussion and concerns expressed by the teachers related to the pending school camp or the small group of students within the grade 8 and 9 classes whose behavior created serious management issues.

School Administration

Issues related to school administration created barriers to the implementation of the integrated project. The first barrier that became evident was related to the subject timetable and room schedule. The school had a custom-built middle school that was designed to enable classes to work in large open areas. Due to timetabling restrictions, however, the grade 8 science classes were not able to work in the middle school rooms and Ms Potter taught her classes in a science laboratory, away from her learning team. As a consequence, Ms Potter had few opportunities to integrate her teaching program with the other subject teachers' programs. The administration agreed to provide Ms Potter with a substitute teacher so that she could participate in meetings with Mr Birchwood, the team leader and social studies teacher, to discuss the project. However, Ms Potter had to arrange these meetings herself and provide a lesson plan and materials for the substitute teacher.

The school administration was responsible for dealing with misbehaving students and providing support for teachers in the classroom. The following vignette provides a reflective account of my field notes taken during Lesson 8 (see Table 8.1) in Ms Potter's science class. During this lesson, one of her students, Garth, became angry and uncontrollable. Ms Potter seemed to be at a loss as she watched Garth interrupt the class, shouting and opening the classroom door.

> It is raining today. The students are waiting outside, lined up against the window trying to keep out of the rain; several of the boys are drenched. Ms Potter is running late, having to come from the other side of the school. The students chat loudly and jostle as they are impatient to get out of the rain. The class troops inside noisily and students push and hustle with their school bags trying to ensure they get to a place next to their friends. Ms Potter sends them back outside to line up again quietly. They line up outside and come back into the classroom muttering. Ms Potter asks the students to get their pens and pencils out of their bags as she hands out the exercise books she keeps in the classroom. Six of the students, mainly boys, have nothing in their bags to write with. Some move around the room and try to borrow a pen. It is now 20 minutes into the lesson. Ms Potter asks the students four questions, which she writes on the board: "1) What is a force? 2) List ten machines you use all the time. 3) Is a fork a machine? 4) Write a definition for a machine." This is revision. Ms Potter later said she revisited the ideas as her other class had been given a short test and did not know the answers to these questions.
>
> Ms Potter asks the class to talk in their groups and then write answers to the questions. For the next ten minutes the students chat in their groups. In the group nearest to me, Helen and Sarah converse about some event the night before and Helen is putting Sarah's hair up. Orion is listening to his iPod and has not yet put the title on his page. Ms Potter has not noticed the leads connecting the noise in his ears to his bag at his feet. The other boy in the group, Gerry, has learning difficulties and has not started writing anything either; he is talking across the room to another boy in the corner and finds it difficult to sit down. Garth has taken a hat from a student and hidden it in another student's bag. Ms Potter stops the class to rein in the escalating noise

and warns the students to behave. Although the noise subsides, it does not cease. Ms Potter dictates the definition of a simple machine to the students three times slowly to ensure all students have written it down. Ms Potter asks the class, "What is a simple machine?" and then gives examples to the class. The noise is escalating and one student has to be moved from his group to the side of the room. Ms Potter gives out the textbook and asks the students to turn to page 15. Students are required to write down the six basic types of machine from the textbook.

At this point, Garth is sent out of the room as he continues to distract other students. Ms Potter takes him into the computer room, which is attached to the science classroom, and sets him up at a small desk with the textbook and paper. It is now 40 minutes into the lesson. Ms Potter returns to class and moves a third student to sit at the side of the room. The students must now copy the notes from page 15 and 16 of the textbook on how simple machines work. Helen complains loudly, "Science is really boring. Why do we have to do such boring science?" Ms Potter warns the class that if they do not work in silence for the remainder of the lesson, ten minutes, she will keep them in at recess. It is now 50 minutes into the lesson. The class settles into silence to complete the note taking. Five minutes before the end of the lesson Ms Potter explains the terms "force multiplier" and "speed advantage/disadvantage," and she verbally gives the students examples. Garth doesn't like being isolated from the class and keeps appearing, opening and loudly closing the classroom door, and waving at the students. Ms Potter ignores him, and he continues to clown around at the door, distracting many of the boys. Ms Potter previously told me that Garth has anger management issues and, as a consequence, is subject to violent and uncontrolled rage. Students are now required to draw a picture and write a few words to explain the terms force multiplier and speed advantage/disadvantage. Helen and Sarah exchange looks and sniggers. Orion still has his earphones in and his music on. Gerry has hardly written anything at all and is still fidgeting in his seat. When the lesson draws to a close the students remain for an extra five minutes before Ms Potter releases them to recess. Through the window they silently watch the other students outside. It continues to rain heavily.

A further school administrative issue was the expected delivery of ten laptop computers that would enable the students to program their robots during Lesson 19 in Week 8 of the term (see Table 8.1). The laptops were due to arrive in Week 6 and Ms Potter was very excited as this would enable all students to program and transmit signals from the computer to their robot. Until that time, the students had used Ms Potter's pre-written program on her personal laptop and, one by one, came to the front of the room to program their robot. The school computer laboratory was unavailable because it had been booked by other classes. Unfortunately, the laptop computers did not arrive during the term of the project and the science department was still waiting for them three weeks into the following term. The non-arrival of the laptops severely impacted on Ms Potter's teaching program and the perceived success of the integrated project.

Teacher Inexperience

Ms Potter was in only her second year since qualifying as a teacher and was finding aspects of the job challenging, particularly classroom management. Her lack of experience of teaching this particular subject was evident; she did not have a well-developed bank of resources and sometimes she was unable to anticipate what she would need for practical activities. This became obvious when the students constructed and sought to program the Lego robots. Ms Potter initially did not have the necessary skills to program the robots herself. Fortunately, several boys who were part of an after-school robotics club helped her during class to write and upload the programs into the robots.

Ms Potter recognized that her inexperience impacted on her skills in the classroom and looked to Mr Birchwood for guidance and ideas. She reported, "he is an excellent teacher and it is really good for me to be able to work with him as I am not very experienced." She referred to Mr Birchwood as her "mentor." She said that students had a lot of respect for him and she felt that if she worked with him that some of the respect "would rub off on her."

Teacher Isolation

Being part of a middle school learning team was isolating for Ms Potter because she was "cut off" from her science peers. The high school science teachers shared an office in the science building that enabled them to have an interactive and collaborative relationship, but this was not available to the middle school teachers who had desks in the middle school learning team offices. As a consequence, there was little opportunity for Ms Potter to plan and discuss her science program with other science specialists. The other science teachers may have had more experience in making the connections between the subject areas and may have been able to suggest and share classroom strategies relating to the topic of simple machines that could link to the project "Making a Difference—Community Access for People with Disabilities." Ms Potter did have one helpful conversation with another science teacher who ran the after-school robotics club. He was able to provide her with some background knowledge to help her get the robots programmed and working.

Science meetings at the school involved a mix of middle and high school teachers and were the only time when teachers teaching science in the middle school were able to talk with other science teachers. Ms Potter reported that these meetings were held infrequently during the term. Moreover, the head of the science department was only able to support Ms Potter in a limited way with behavior management. When Ms Potter had badly behaved students when teaching in the science department, the head sometimes removed students and isolated them in other classes.

Disaffected Students

The behavior of the students in Ms Potter's classroom impacted on her choice of activities, and often the behaviors of a few individuals impacted on the learning of the entire class (see the vignette included above). This was evident when the teacher was forced to repeat material in a subsequent class due to poor behavior (see Table 8.1), and had to stop the students working on the Lego robots and require them to write notes from their textbooks. Another problem faced by the teacher was the poor attendance rate in both the grade 8 classes. Often up to a third of the class was missing on any given day. Although the smaller classes were easier to teach, when students returned they did not understand what the class was doing and it was difficult for them to catch up. At other times, students were not prepared for class and Ms Potter had to remove students from the class due to disruptive behavior.

Although it was predominantly the male students who misbehaved in the classroom, a few of the female students were vocally dissenting toward the project. As noted in the vignette, one student complained loudly to the teacher that it was really boring and she would rather be doing something else. Ms Potter sat this student outside of the room with a textbook and some questions to copy.

At the end of term, several students were asked about their experiences and whether they had enjoyed the project. One student responded, "I just thought it was all boring, like everything we did. And so the whole term was boring and everything. Nothing really interested me in it." When this female student was questioned about whether she found learning about people with disabilities interesting, she responded: "We didn't. I don't remember doing that. No, not in science. In social studies I did." When the researcher asked her if she enjoyed learning about people with disabilities in social studies, the student responded that she did.

The data collected from the survey and interviews indicated that many students, like the student in the previous paragraph, did not make the connections between the science and social studies content through the "Making a Difference" project. The aspects of the project that related to science did not seem to engage the students, although the social studies component of the project did seem to interest them. Despite the aims of the project, students rarely saw the science aspects as relevant to their future, or useful in their everyday lives. This was not the case for all interviewed students, however, with one student reporting that he could see clear links between the subject areas, and also links to people with disabilities.

While several students displayed considerably disturbing behavior, Ms Potter demonstrated that she sincerely cared for her students. For example, when one student, Garth, who had previously been disruptive, completed a set task to a high standard, Ms Potter's delight was clearly evident. It was all she could discuss with me and other learning team members during lunch that day.

Final Reflections from the Teacher

From Ms Potter's perspective, this experience was "an extremely steep learning curve." She conceded that there were aspects of the project that she felt were valuable and others that she thought were too difficult and, if the project was repeated, she would not do again. Overall, however, she felt that the "Making a Difference" project was successful because the students had an increased awareness of people with disabilities. She saw the science and social studies subject components of the project as complementary. According to Ms Potter, the science helped the students to identify problems and find solutions, and social studies enabled the students to communicate the issues to the broader community, and gave them a more humanistic perspective.

Ms Potter mused that, although a number of students really enjoyed working on the robots, she would not use the robots again as it was difficult to keep track of all the Lego pieces. She also felt that some of the students struggled with the concept of integration and were not easily able to make the necessary links between the content because they considered the subjects to be unrelated to each other. She also was disappointed, but accepted, that some of the other teachers in the learning team did not want to participate in a collegial way.

Ms Potter's delight when students experienced success was palpable but her distress when students did not try, behaved badly, and needed to be removed from the class, affected her a great deal. Her emotions changed noticeably, sharing the students' joy and their disasters. During the final excursion (see Table 8.1), when the students and teachers got to race each other in wheelchairs at the Wheel Cats Basketball Courts, her enthusiasm and enjoyment of teaching and of the students was visible in the excitement that they shared as they raced down the court. Ms Potter said to another teacher, "This makes it all worth it. We are having as much fun as the kids."

Conclusion

This chapter highlighted the frustrations and barriers an inexperienced teacher faced while seeking to integrate science with other subjects including social studies through a community-based project in a specialized middle school. The barriers included lack of collegial collaboration from teachers of other subjects, poor support from the school administration, the teacher's own inexperience and her isolation from other science teachers, and the poor behavior and lack of engagement of the students in the topic. Despite these barriers and challenges, the teacher thought that, overall, the project was a success because of the valuable and contrasting insights that the subjects of science and social studies brought to the topic of understanding people with disabilities.

Focus Questions

1. The learning team coordinator delegated responsibility for the science part of this integrated project to Ms Potter. Given the challenges she faced that are described in this chapter, do you think this was an appropriate action? What benefits may have arisen for Ms Potter?
2. How confident are you in your knowledge and skills to teach a topic such as the one described in this case? What are some of the difficulties that might arise in teaching topics involving socio-scientific issues? How might these difficulties be overcome or forestalled?
3. What are the professional obligations of more experienced teachers and school leaders in providing support for newly qualified teachers? What concrete actions can be taken to support inexperienced teachers, particularly when they are endeavoring to introduce new curricula?

Suggestions for Further Reading

Levinson, R. (2006). 'Towards a theoretical framework for teaching controversial socio-scientific issues'. *International Journal of Science Education*, 28(10), 1201–1224.

This article develops a model for teaching controversial socio-scientific issues to middle and high school students. The model includes three categories of: reasonable disagreement, the communicative virtues, and modes of thought. Examples in contexts such as genetic diagnosis, climate change, and nuclear power are given to illustrate how the model can be used by teachers so that the controversial features are made explicit to students.

Burrill, J. & Hernández-Gantes, V. M. (2003). 'Team planning to create an integrated curriculum'. In S. A. McGraw (Ed.), *Integrated mathematics: Choices and challenges*. Reston, VA: The National Council of Teachers of Mathematics, Inc.

This chapter focuses on collaboration to enable the planning and delivery of an integrated curriculum. The authors discuss the issues surrounding selecting and building a team, team planning, designing worthwhile problems, and facilitating collaboration.

References

American Association for the Advancement of Science. (1993). *Benchmarks for scientific literacy*. New York: Oxford University Press.

Lewis, J. & Leach, J. (2006). 'Reasoning about socio-scientific issues in the science classroom: The role of scientific knowledge'. *International Journal of Science Education*, 28(11), 1267–1288.

National Research Council. (1996). *National science education standards*. Washington, DC: National Academy Press.

——(2011). *A framework for K-12 science education: Practices, cross-cutting concepts, and key ideas*. Washington, DC: National Academy Press.

9

FOCUS ON LEADERSHIP

Constructing a Model House at Mossburn School

Rachel Sheffield

Introduction

Curriculum innovations involving several teachers working together across different classrooms entails delicate campaigning, maneuvering, and negotiation (Hall & Hord, 1987). Integrated curricula pose particular challenges because they often require teachers with different subject expertise to work together across disciplinary boundaries. Sometimes this collaboration takes place in teaching teams that are already established in a school, particularly at the middle school level. Other times, such collaboration involves teachers from separate subject departments who rarely work together. However, we do know that teacher collaboration is an important contributor to teacher learning. Collaboration provides the means through which innovative ideas and approaches can be tested; it enables information and insights to be shared, and it also enables common issues to be discussed and debated (O'Donoghue & Clarke, 2010).

The process of implementing a collaborative, integrated project also requires leadership—a person or persons with the vision to guide others to work toward a common curricular goal. Sometimes, the leadership is shared, whereby people lead and learn together, and construct "meaning and knowledge collectively and collaboratively" (O'Donoghue & Clarke, 2010, p. 158).

This case story examines the tensions experienced by the project leader and other grade 8 class teachers when they tried to implement a model house construction project involving several subjects, including science, English, religion, Arabic language, social studies, and mathematics. In this instance, the leader of this project was not one of the classroom teachers but rather acted in the role of coordinator. These circumstances complicated the leadership role and made it more challenging than it would have been had the project leader been a regular member of the teaching team. We found that, while the project was successful in integrating some subject areas, there were also several challenges and tensions that needed to be recognized and resolved.

Mossburn School

Mossburn is a small, religious, independent school with students from grade 1 to 12, many of whom travel considerable distances in the school's private buses to attend each day. In the year of our research, there were six grade 8 classes (students aged 12 to 13 years), one of which was the academic extension class that participated in the model house construction project which was the focus of this case story. The academic extension class was different from other grade 8 classes because it was the only class that included both boys and girls in a co-educational environment. Other classes were comprised of either boys or girls only. The 30 students (8 girls and 22 boys) in the academic extension class were very focused and hard working. They understood that being in the extension class was a privilege that could be revoked depending on their achieved grades and level of engagement as reported by their teachers. The number of students in the academic extension class did fluctuate throughout the term of the case story with one student leaving the class and two others arriving. The students participated in the model house construction project in addition to their normal classroom activities in each of their subject areas. No extra time was given to the students and teachers to conduct the project; instead, time was found by condensing other content. The academic extension class, known within the school as Grade 8 Yellow, remained in their home room for all their learning sessions with the subject teachers coming to their room. At the end of the term, all grade 8 students, including the academic extension class, participated in school-wide testing for all subjects.

The Mossburn School curriculum was based on the state guidelines and the participating students studied English, mathematics, science, social studies, physical education, computing, Arabic language, and religious education. The students did not study design or engineering subjects and, as a consequence, had no access to a trades workshop and were not taught the practical skills for building a model house. Within each subject, the related teaching program for the project focused on a number of conceptual areas, for example, in science the students were studying the broad conceptual area of energy with a focus on electricity.

Case Story: the Model House Construction Project

Project Leadership

The model house construction project was introduced two years previously in a limited capacity with two teachers working together to integrate mathematics, computing, and science in the grade 8 academic extension class. Ms Rahim taught science and Ms Haslam taught the mathematics and computing. At the end of the project in each of these two years, the students' models were displayed at a parent night at the school, thus providing a public acknowledgement of students' work. The project was expanded in its third year to integrate a larger number of subjects and involve more teachers. Although Ms Rahim did not teach the grade 8 academic extension class in the third year, the year of our research, she became the project coordinator and was instrumental in drawing together the teachers in meetings throughout the term. Ms Rahim's role was to help the teachers with the integration of activities and, at the same time, ensure they were meeting the assessment objectives for their respective subjects. The science related to the model house construction project was taught by a different teacher, Mr Caspian. From our prior experience with integrated projects, it was unusual for a project to be directed by a person who was neither a member of the learning team nor the administration.

Staff Meetings

Our first meeting with Ms Rahim occurred in the term before the project started. Ms Rahim said she was considering how to shape the project as it entered its third year, and was hoping to expand it from a few subjects into a larger project across a wider range of subjects. Our application to the principal for permission to visit the school and observe the project in action possibly resulted in the project having more impetus than it may otherwise have had, encouraging Ms Rahim to extend the project. This is one of the interesting side effects of observing educational projects and engaging in collaborative research with schools; we are never sure how the project would have developed without our participation, even as non-participant observers. It is possible that the principal at Mossburn School considered that the involvement of our research team would be beneficial for the teachers and students, and he also may have been hoping for a positive public relations outcome through the research team's participation.

One of the early meetings of the Grade 8 Yellow class teachers was held during class time with substitute teachers provided to ensure that all teachers were available. However, the science teacher, Mr Caspian, was unable to attend due to other in-school commitments. As she had taught the science topic previously, Ms Rahim represented Mr Caspian as the science spokesperson.

Ms Rahim developed and distributed an agenda to the other teachers before the meeting so they could check how their aspect of the model house construction project had been conceptualized as part of the larger project. The agenda described how the project would be introduced in the different subjects and presented a tentative outline of the teaching and learning tasks to be completed. In some subjects, the assessment rubrics were already designed and were presented during the meeting. This enabled all teachers involved to see how the other teachers would approach the model house construction project, the major concepts to be covered, when the assessment was scheduled, and the marking rubric.

TABLE 9.1 Student Learning Tasks Planned for Each Subject in the Model House Construction Project

Subject	Teacher	Student tasks for the model house construction project
Science	Mr Caspian	Design, build, and paint a model house of balsa wood and cardboard. When built, install a working parallel circuit of lights with switches.
English	Ms McDonald	Write a report using the data collected from other subjects with a focus on the process of report writing as presented in a poster.
Arabic language	Ms Hassan	Complete one of the following two tasks: 1. Draw and label a house using Arabic language terms; 2. (For more able students) Write a newspaper advertisement in Arabic language describing a house for sale.
Religious education	Ms Haslam	Prepare a talk about the factors that make a happy, Islamic home considering how Islamic values are enacted in homes.
Computing	Ms Sawyer	Use the computing program "Sketchup" to design and create the floor plan and external features of the model house.
Mathematics	Mr Samei	Draw a scale drawing of the model house and connect to the local city council website to determine the utility rates.
Social studies	Mr Hashim	Research the history of houses and how housing has changed over the ages.
Physical education	Mrs Hanwell	Prepare a five-week eating and exercise plan that promotes healthy living.

The project comprised a series of interrelated subject tasks rather than an individual, holistic project (see Table 9.1). One document was provided to the students that contained a cover page, the project design brief, a calendar for the term, details of the tasks in each of the subjects, and key assessment dates. As can be seen from the learning activities listed in Table 9.1, some teachers were able to make more plausible links to the main task of building a model house than others. For example, the computing teacher, Mrs Sawyer, said she was really pleased about the collaboration because she wanted to use the "Sketchup" software program with the students and the model house construction project provided an authentic context for students to learn how to use the program. The physical education teacher, Mrs Hanwell, however, found it difficult to make explicit links between her task, to prepare a diet and exercise plan, and the model house project.

Ms Rahim appreciated that, at Mossburn, the curriculum was assessment-driven and class time was limited, so she tried to ease the teachers' load by focusing on the marking rubric before the start of the project. She wanted to make sure that teachers had plenty of time to consider how they could develop their tasks to be compatible with both the project and the common subject assessment at the end of term. However, this created a dilemma for the mathematics teacher, Mr Samei, because Ms Rahim's expectations of a project-related assessment task for the students was not compatible with the common, grade 8 test administered by the school's mathematics department. Mr Samei said that he needed to be seen to be doing the "right thing" for the project, which was supported by the principal, but he did not feel that the project could be integrated with the current grade 8 mathematics curriculum. Mr Samei did not provide his assessment tasks or rubrics prior to the staff meeting and Ms Rahim subsequently commented that she had to follow up several times during the following weeks to get the necessary material. The mathematics component of the model house building project was, therefore, an addition to the mathematics learning objectives set for the term by the mathematics department.

Part of the teachers' meeting was used to help inform participating teachers what students would be completing in their other subjects. Toward the end of the staff meeting, the social studies, English, and religious education teachers became engaged in a lively debate about the authentic embedding of values in the various subjects, including the many commonalities between social studies, the environment, and science. The social studies teacher, Mr Hashim, expressed his view that the value statements from the state curriculum clearly linked the content that can be taught in the subjects of science and social studies. Ms Rahim encouraged the expression of ideas and there was discussion about whether values could be included implicitly or explicitly as part of the model house construction project.

Classroom Observations

As part of our research program, we visited the school to attend staff meetings and observe the teaching of subjects connected with the model house construction project. All science lessons were observed, and also some lessons in mathematics, social studies, and a computing class that dealt with topics related to the house project. Each class was taught in the same classroom, and, while this created a sense of continuity, it was sometimes difficult to see the connection and links being made by each subject teacher to the model house construction project. The following sections of the chapter provide summaries based on classroom observations, informal conversations, and more formal interviews with the participating teachers.

Science

The project director, Ms Rahim, accompanied us to most science lessons and nearly always stayed to help the students set up necessary equipment and to discuss the project. The students created their houses from balsa wood and, as the school had no design or engineering workshops, the materials and tools were supplied by the school maintenance staff. The students completed most of the house construction during science, and clearly enjoyed the process of planning and building the house. They worked in groups, negotiating tasks and activities within the group. The following paragraphs provide a summary of a set of field notes made during a visit to Mr Caspian's science class near the beginning of the model house construction project.

> When I arrived at the school I was "buzzed in" from behind a locked glass window and signed in at the front office before being collected and escorted through the school by Ms Rahim. We chatted as we made our way up the stairs to the classroom of the grade 8 academic extension class. Ms Rahim expressed her excitement that the project was underway and the students seemed to be interested and engaged by the notion of designing and constructing an actual model house with lighting.
>
> Ms Rahim stayed with the class during Mr Caspian's science lesson, helping with the implementation of the project. As a result, there were two teachers and one researcher in the small classroom with 30 students. The eight computers on computer desks around the room gave a cramped impression. Early in this lesson, the balsa wood and tools that would be used to start building the model houses for the project were delivered by several of the school's maintenance staff. There were limited supplies as the tools came from the maintenance shed, there being no design or engineering program at the school.
>
> The students had already formed groups; the boys and girls worked in separate groups on different sides of the classroom. Plans for the model house had already been drawn on paper, so students first had to transfer the floor plan of their model house on to the balsa wood base and then use saws to cut the wood to the correct shape and size. Ms Rahim set up the sawing of the wood on the verandah and a queue formed as there were not enough saws for all groups to use at the same time. The students enjoyed the process of drafting and then cutting out the materials, although many students were unfamiliar with the tools. As a consequence, there were several bloodied fingers, but cuts were minor and shown to peers with pride, particularly by the boys in the class. Mr Caspian did not come out on to the verandah to oversee the students in his class while they worked on their model houses. He remained inside the classroom behind his desk while Ms Rahim helped and supervised the students on the verandah.
>
> Mr Caspian did not talk to the class for the remainder of the lesson; however, he spoke with individual students when they went to his desk to show him their construction or ask for his opinion or advice. Toward the end of the lesson, it was Ms Rahim who called the students to order and started them collecting and returning equipment, and putting off-cuts of wood into the trash. At the end of the lesson, it was Ms Rahim who dismissed the class before escorting me back to the school office so I could sign out of the

school. She was very enthusiastic and chatted happily about how much the students seemed to be enjoying the construction process and what other aspects of the project they had been pursuing in their other subjects.

We noticed that between our visits, which sometimes spanned as much as a week, very little was done on the house construction. We speculated that this was due to two factors. The first factor was that the students had very little experience working manually with equipment and also needed to negotiate tasks within their groups. The second factor was that the science teacher, Mr Caspian, used the lessons to focus on the common objectives of the energy topic that would be tested at the end of the term. As a consequence, students were unable to complete their houses before the end of the term due to insufficient time and slow progress. The students were not ready to show their projects on the parent–teacher night as had been the case in previous projects. Unfortunately, this prevented a rewarding closure or culmination of the project, and public acknowledgement and celebration of the students' hard work was not forthcoming.

Mr Caspian appeared to have little interest in, or enthusiasm for, the model house construction project. Observations of his teaching of electricity showed a strong focus on "chalk and talk," with students quietly taking notes. He made no comment to us about Ms Rahim's presence in the science classroom; however, he later commented in an interview that a drawback of the project was a lack of effective organization. He stated, "It is important to integrate subjects which have similarities and undue integration is not very beneficial." Although the science component was pivotal to the project, he perceived a lack of cohesion and organization. These observations suggest the project was not encouraging distributed or shared leadership, but had resulted in a more autocratic style of leadership from a person external to the teaching team whose enthusiasm and commitment was the driving force for the project.

Mathematics

During one observed mathematics class, Mr Samei tried to get the students working on the computers to connect to the local city council website to determine the rates for utilities, such as electricity and water, which would apply to the house. He also wanted the students to research the local building rules and regulations, and how these might vary between different city jurisdictions. Students were required to use the school computers to search the local government websites for relevant information; however, this proved impossible in class due to firewalls on the school server, which severely limited students' online access. The failure to connect to the necessary websites in class led to the students becoming confused and frustrated. As a consequence of these frustrations, Mr Samei relegated the mathematics component of the model house construction project to be completed for homework and, in this and subsequent lessons, he went back to following the common school mathematics curriculum. Through these actions, it seemed to us that Mr Samei felt little loyalty to the model house construction project and completing the mathematics part of it was low on his agenda. The main focus for him was being able to complete the mathematics curriculum as set by the mathematics department.

English

The English teacher, Ms McDonald, demonstrated a strong interest in using the skills and processes that students were learning in English to bring together the other learning areas through the model house construction project. Her students set up a journal so they could write about, and reflect on, the model house project. Ms Rahim reported that this journal writing resulted in unexpected benefits for students, who had enjoyed the reflective component. Ms McDonald also encouraged the students to create posters that related to the design brief of the project and these were then displayed on the walls in the classroom. Ms McDonald allowed the students to work on their houses during her English class time, but only on limited occasions.

In contrast with her seemingly highly supportive actions in class, comments in Ms McDonald's post-project interview revealed frustration at the timing and collaboration, and some discontent with regard to the nature of the project. For example, she stated "[I] did not have enough time to complete [the house project]. I needed two terms. Staff need to be given more time to liaise throughout the project." Ms McDonald also noted that, "this [project] does not really suit the academic focus of the school."

Social Studies

The social studies teacher, Mr Hashim, was similarly keen to incorporate the model house construction project into the social studies lessons. He expressed his interest in being involved in the project and he was very enthusiastic in discussing the overlap of values during the staff meeting. Mr Hashim commented in an interview after the project that, "the cooperation between teachers was excellent, in terms of trading class time, etc." He made considerable effort to integrate the model house project in an authentic way into the social studies curriculum while also meeting the common subject outcomes. Mr Hashim and the religious education teacher, Ms Haslam, seemed to be the strongest supporters of the project. Ms Haslam had considerable flexibility in her subject curriculum to incorporate social and family aspects of the model house project into the classroom activities.

Students' Responses to the Project

The students responded positively when asked about their project experiences and we had lively, engaging debates with the students at the conclusion of the project.

In the final lesson for the term, students arrived in groups and were very excited about showing off their model houses and explaining and discussing their achievements and the challenges involved. The student interviews conveyed their knowledge and enthusiasm about their model houses and how they used knowledge from multiple sources during the planning and construction. For example, in one interview, students provided a detailed explanation of how they researched the function of double-glazed windows in their science lessons.

During the post-project interviews students explained who had helped them with their project. For example, "I talked to my mom and dad, and my dad helped me with the angle of the roof, and the shape and design" (Student L). Another student explained, "My dad's friend is a builder, so we always go and visit him. So he just helped me with some of the building regulations and stuff" (Student F).

Ms Rahim had hoped that the products and other outcomes of this project would be showcased to the wider school community. She said that in previous years the enthusiasm, pride, and engagement of the students and their parents were very positive. Ms Rahim explained that

> Some of the parents were judges, and went around and checked all the houses and the students had a chance to talk about their houses, show us their house plans, why their houses were so great.

For this year, however, the students did not have sufficient time to complete the model house construction project, and the projects were sent home at the end of the term unfinished. The models were clearly in need of further time and effort, especially for the girls who had planned to finish their model houses with flowers and fences. The students reported their disappointment at not having completed their houses to their satisfaction.

> This, our front yard, is where we're gonna [sic] paint it green and put little flowers, you know, decorations. (Student S)

Leadership Tensions

Tension surrounding aspects of leadership became evident during the implementation of the project. As the project progressed, it soon became evident that several teachers did not contribute to, or share responsibility for, the integrated project in ways that reflected their initial commitment. Some teachers were less interested in the aims of the wider project than the impact of the project on their own subject area. Some of this could be related to the busy schedules that the teachers were following. Ms Sawyer, the computing teacher, admitted, "Well, I must confess, because of the workload I did not get a chance to talk to other teachers about how it was going, but, as far as computing was concerned, it started on time and finished on time." Ms Rahim focused on ensuring that all teachers kept to the schedule they had set and submitted marks for assessment items that they had contributed to the project brief, thus ensuring that teachers did make a contribution to the project.

Ms Rahim and Ms Haslam had previously run the project very successfully. To Ms Rahim, the focus of integrated learning was to link the subjects with students' experiences so they could see more relevance in their learning. She remarked that

> the students usually talk to us, saying things like: "Yeah, but why am I doing this? It's so irrelevant." But [with integrated topics] you make it more linked so they feel that it's worthwhile learning.

As Ms Rahim reported:

> We had a science show at the end of it and the parents were invited to come and so was the principal, and they were the judges as well.

This perceived success may have resulted in greater pressure on the teachers to become involved in the project in the following year. On the one hand, the fact that Ms Rahim had previously managed the project so successfully may have inspired confidence among the teachers involved in the project. On the other hand, this fact may have had a negative impact, because the project was perceived to be "Ms Rahim's project" with few teachers having a sense of ownership. Some teachers did not make either positive or negative comments, but it seemed that they just wanted to complete their aspect of the project and assess the students' work. For example, this can be seen in Ms Sawyer's comment that the success of her component was indicated by the fact that it started and finished on time. This belies the notion of shared leadership. Paradoxically, it is possible that the previous success of the project, implemented by Ms Rahim and Ms Haslam, contributed to the lack of ownership among this larger group of teachers.

Conclusion

There is growing evidence that an integrated approach to schooling can have a useful role in preparing young people for future engagement with the social consequences of scientific developments. At the same time, research has identified that long-term, sustained integrated approaches are extremely difficult to maintain and often ebb away with time (Meier, Nicol, & Cobbs, 1998; Wallace, Sheffield, Rennie, & Venville, 2007). This case story delves into the leadership complexities of implementing a curriculum that attempts to draw connections between subjects that are usually taught and assessed independently.

Leadership is an important attribute and it is evident that some degree of shared leadership is necessary for collaborative, integrated projects so that they result in positive teaching and learning outcomes. This case story shows that, although the presence of a strong leader was helpful in the implementation of the project, some teachers in the learning community did not share responsibility for implementing their part of the project and, as a consequence, the project was not as successful as it might have been. Based on our observations, we make the assertion that leadership is a pervading attribute that can potentially impact other attributes that contribute to the successful implementation of an integrated project.

Focus Questions

1. What kind of challenges do you think you would face if you wished to lead a collaborative, integrated teaching and learning project? How might you prevent or overcome these challenges?
2. In this project, some teachers were unable to see connections between their subject and the planned model house construction project. How can we find an appropriate balance between the curriculum requirements of an integrated project and traditional subject matter?
3. What strategies could be used to encourage a reluctant team of teachers to become involved in an integrated project in a positive way?

Suggestions for Further Reading

O'Donoghue, T. & Clarke, S. (2010). *Leading learning: Process, themes and issues in international contexts*. Milton Park, Abingdon, UK: Routledge.

The authors of this book promote the view that stakeholders in school transformation have complementary leadership roles they must play to enhance student learning. They discuss the notion of distributed leadership and the importance of teacher collaboration to the process of school improvement and the professional learning of teachers themselves.

White, G., Russell, T., & Gunstone, R. F. (2002). 'Curriculum change'. In J. Wallace & W. Louden (Eds.), *Dilemmas of science teaching: Perspectives on problems of practice* (pp. 231–244). London: RoutledgeFalmer.

This chapter explores curriculum change and the circumstances surrounding teachers when they are caught between a desire to change and improve their practice, and pressure to conform to long-established patterns of classroom organization and student learning. It is about balancing the risks and the rewards of change.

References

Hall, G. E. & Hord, S. M. (1987). *Change in schools facilitating the process*. New York: State of New York Press.

Meier, S., Nicol, M., & Cobbs, G. (1998). 'Potential benefits and barriers to integration'. *School Science and Mathematics*, 98(8), 438–447.

O'Donoghue, T. & Clarke, S. (2010). *Leading learning: Process, themes and issues in international contexts*. Milton Park, Abingdon, UK: Routledge.

Wallace, J., Sheffield, R., Rennie, L., & Venville, G. (2007). 'Looking back, looking forward: Re-searching the conditions for integration in the middle years of schooling'. *Australian Educational Researcher*, 34(2), 29–49.

10

FOCUS ON COMMUNITY

Learning About Tiger Snakes at Chelsea Elementary School

Rekha B. Koul and Rosemary Sian Evans

Introduction

In this chapter we discuss our findings from an integrated, environmental science project, titled "Living with Tiger Snakes" (LWTS), which was based at an urban wetland that is a habitat for venomous tiger snakes (*Notechis scutatus*). Tiger snakes are commonly sighted around the wetlands, and the wildlife center located at the lake receives hundreds of calls each year from concerned members of the local community who find tiger snakes on their property. The manager of the wildlife center, Mr Roberts, is able to arrange for the snakes to be "relocated," but, due to fear and ignorance, snakes are often killed by property owners.

Mr Roberts was keen to conduct an education program to increase community awareness of the snakes, discuss appropriate safety measures, and outline the snakes' importance in the wetland ecosystem. He applied for, and secured, funding for a joint education program with Chelsea Elementary School, located nearby. Ms Edwards, a lead teacher at the school, believed that the program offered an opportunity for students to learn about environmental responsibility, biodiversity, and wetland ecology. The funding for the project came from a school, community, industry partnership in science program, and, although modest at $3000, was able to provide payment to the center for Mr Roberts's time (the wildlife center was required to generate some of its funding) and to transport students between the school and the wetlands, a 15-minute bus journey.

In the following sections, we explore some of the issues that arose during the LWTS program, including the cooperation between the wildlife center and the school, the involvement of the community, and the extent to which environmental science was integrated with other subjects in the school curriculum.

Chelsea Elementary School

Chelsea is a small independent elementary school with only five classes, a kindergarten class, two classes of combined grades 1 to 3 (children aged 5 to 8 years), and two of combined grades 4 to 7 (9 to 12 years). All 45 students in grades 4 to 7 were involved in the project. These students were taught by two classroom teachers, Ms Jackson and Ms Mariott. As lead teacher, Ms Edwards was responsible for the school's science program (among other responsibilities) and frequently team-taught with the other teachers. She led the school's activities relating to the LWTS project, and was the person who liaised with Mr Roberts for the duration of the program.

The LWTS project was spread over seven weeks, beginning with two full-day visits to the wildlife center, followed by shorter sessions at school or the center of two to four hours. The program culminated with an evening event at the center, to which parents and other community members were invited. Here, every child was involved in a presentation of some kind to portray the work they had done and the knowledge they had gained. We attended all the formal lessons addressed to the LWTS program at Chelsea and the wildlife center, and also attended the community night at the end of the program. We independently recorded our observations as field notes during the students' four visits to the center, which included two activity days, one rehearsal afternoon, and the community night. Observations were also recorded on three occasions at Chelsea when students collated their community survey data and during an activity afternoon with Mr Roberts. The observation data from each researcher were compiled to produce a single set of detailed field notes. The opinions of the three teachers, Mr Roberts, and the students about the program were established through casual conversations. A similar approach was adopted with parents and students on the community night to explore the perceived effectiveness of the project. Formal interviews were conducted with Ms Edwards, Ms Jackson, Ms Mariott, and Mr Roberts at the end of the project.

Case Story: LWTS

The formal work of the program was conducted on one day in each of seven weeks. Table 10.1 gives an overview of the kinds of activities undertaken during the researchers' visits to the center or to Chelsea. However, Ms Edwards and the class teachers informed us that much additional time was spent at school, particularly in preparing the students' presentation at the community night.

Implementing the LWTS Program

The activities at the wildlife center were planned and led by Mr Roberts, the center manager, and all of the students and teachers participated. On the first day at the center, Mr Roberts introduced the program and set out the expectations for the outcomes, which included students conducting a survey of parents and neighbors to

TABLE 10.1 Description of the Timeline and Activities of the *Living with Tiger Snakes* Program

Week	Venue	Activity	Time
1	Wildlife center	Introduction to tiger snakes by herpetologist, handling python First aid for snake bite demonstration and practice Snake walk and observation Begin preparation of a community survey to determine community awareness of snakes Planning posters and future events	full day (including travel)
2	Wildlife center	Introduction and nature walk Working on posters Demonstration about nest boxes Dipping in the lake and learning about the wetland ecosystems	full day (including travel)
3	Chelsea	Coding and collating survey data collected by children out of school time	half day
4	Chelsea	Lesson on food pyramids and food chains by Mr Roberts Food web activity led by Mr Roberts	1.5 hr
5	Chelsea	Preparation of presentations for community night	various times during week
6	Wildlife center	Rehearsal for the community night, organized by teachers	2 hr
7	Wildlife center	Community night—Introduction and snake exhibition by expert Student presentations of posters, first aid role plays, etc.	2 hr

establish community attitudes to snakes, preparing posters about snake safety and "carry cards" with instructions for first aid for snake bite as part of the community education, and the show-and-tell activities for the community night. Students listened attentively to a talk by a herpetologist, who showed the students a python they could handle and a caged tiger snake, which, for safety reasons, they could not touch. Students enjoyed a walk around the wetland to look for snakes in their natural habitat. (A few excited and imaginative children were sure they had seen a snake.) After lunch, Mr Roberts proposed that a community survey be administered by the students. Students worked collaboratively with Mr Roberts to develop survey questionnaire items for the survey but did not have time to complete it. Students were shown how to deal with a tiger snake bite emergency and practiced this first aid on each other. To conclude the day, students discussed ideas for posters they could make about what they had learned.

On the second whole-day visit to the center, students had another tour of the wetlands (but again there were no certain snake sightings), worked on their posters and surveys during the morning session, and in the afternoon a wetlands ecosystem discussion was facilitated by Mr Roberts followed by students collecting organisms from the wetlands and examining them. Mr Roberts explained the behavior of these organisms and their function in the ecosystem to the children. Mr Roberts said that he believed that

> being at a real-life venue adds value to the whole experience and also a one-off experience is great. Kids come here enthused and they think it's great, exciting. Action-centered learning is what we propose, we promote here.

Our observations and informal discussions with students suggested that they were fascinated by the snakes and greatly enjoyed practicing first aid on each other after pretending to be suffering from snake bite. Students also expressed interest and surprise at the variety of small organisms they discovered in the water from the lake.

Students worked on completing their surveys at school and administered them to their parents and neighbors at home. They obtained over 190 responses from community members about their understanding of and attitudes toward tiger snakes. The coding and collating of data was completed over two days at the school under teacher guidance.

During Week 4, Mr Roberts visited the school to present a 90-minute session on food chains and food webs, followed by students moving about role-playing elements of the food web. The session was well planned and well received by the students, who particularly enjoyed the role play, and by the teachers, who appreciated Mr Roberts's knowledge of the wetland ecosystem.

High levels of interest continued when students returned to the center for the dress rehearsal for the community night. Students worked in groups of four to prepare PowerPoint presentations of results of the community survey, or to demonstrate models they had made (such as food pyramids, or a diorama of gardens that were snake friendly or not), posters and carry cards (snake safety and first aid), and act out small dramatic plays of snake bites and administering first aid.

The community night was well attended by parents, siblings, and friends. The aim of this program was to increase community awareness about tiger snakes via the students' presentations. Additionally, a snake expert showed several tiger snakes to the audience. This event marked the end of the project and, apart from the student-led community survey, was the main involvement of members of the community. The evening progressed successfully and the audience was very positive about the work students had completed. The snake expert's presentation and his snake exhibit was a particularly successful way to conclude the evening. Parents were appreciative of the project, particularly the enthusiasm of their children, and expressed these views to the researchers.

Outcomes of the LWTS Program

The outcomes of the project are related to the integration of a real-life experience into classroom activities, factors associated with the community partnership, and the integration of values, knowledge, and skills into the school curriculum.

The Integration of a Real-Life Experience

According to Mr Roberts and the three Chelsea teachers, the opportunities provided by the LWTS program resulted in increased motivation and enthusiasm that lasted throughout the entire program. This particular real-life experience focused on the natural interest of the children and involved many contextual activities. Mr Roberts explained in an interview:

> Kids of that age have a curiosity and natural interest in all things natural and we tap into not only that curiosity, but also the fact that … the tiger snake is pretty interesting; it can kill you, and people are scared of it. So you can build a lot of motivation around that. … The other thing is that the issue was a real-life problem. Around here [snakes] are a real issue and these kids being close by would have seen tiger snakes, would have had experiences, or known someone that had snakes in their backyard and didn't know what to do with them … so they know that it is a real issue, and they can identify with that.

Mr Roberts believed "the fact that they [the students] are involved in the problem-solving of a real problem" added to their enthusiasm.

Ms Edwards thought the impetus and enthusiasm of the students for the whole project stemmed from the initial visit to the wildlife center:

> The children enjoyed the wetlands experience enormously. They gained a lot out of it, being in that environment and … it was a huge level of excitement and enthusiasm for the project. … It definitely added the real-life environment, them actually walking through the wildlife center.

Ms Jackson commented that

> Mr Roberts's knowledge had a huge input. He was walking around with the children and telling them about these things that he knew and that we wouldn't probably have any idea about was really valuable. I think it made a difference being somewhere else other than school. I think it made a difference for the kids too, that they were actually presenting or performing in a different venue.

Reflecting on the enthusiasm of students, Mr Roberts commented that the enthusiasm of the students was a tribute to the teachers and the ethos of the school. He believed the teachers were very committed and the school was "beyond the norm." As he explained, the students

> were motivated and you could see that came across ... helping the community, which that school was particularly geared toward, that civic pride and social responsibility.

Partnership between Chelsea and the Wildlife Center

The teachers and Mr Roberts agreed that the project consumed much of their time and effort, but in different ways. For the center manager, it involved a large amount of paperwork at the outset to apply for funding to run the project. Although the plans for the program were clear in Mr Roberts's mind at the beginning of the partnership, the teachers were not aware of the amount of extra work they would be involved in to make sure the outcomes were achieved. Ms Edwards described her initial understanding:

> The way it was presented to [the principal] was that we wouldn't have to do very much, our kids would benefit by going over there, having some great experiences and there wouldn't be a great workload on us at all. That wasn't, I don't think, investigated or understood fully. That to make it work, as it was set out in the submission, there would have to be a lot of work [done by us].

These different understandings between the members of the partnership resulted in the restructuring of the planned science program on "flight" for the grades 4 to 7 students and an increased workload for the students and the teachers at the school. As Ms Jackson explained

> I think initially because our understanding was it wasn't a lot of work to do, that we felt that we could fit it into a few days and this was just going to be tiger snakes, but when we realized the enormity of the project and because we wanted a good result and we thought it was so worthwhile, [we decided] to put a bit of flight aside and to focus on the tiger snakes.

Ms Mariott concurred and added that as the end product of the project was a showcase public performance, she believed it became even more important to ensure time and effort was given to the tasks.

A further issue for the teachers at Chelsea was the difference in teaching styles between Mr Roberts (a trained high school science teacher) and the Chelsea elementary-trained teachers, together with the availability of necessary resources when working at the wildlife center. Ms Mariott considered that students' progress was impeded because the management of 45 students at the center was different from the way the students were usually managed in their homerooms at Chelsea. Ms Mariott thought that, although the students worked well at the center, they did not have access to their regular resources, such as library books and stationery items. Consequently it was necessary for the students to complete a large proportion of their work at school.

On the contrary, Mr Roberts believed that "it would have been nice if we could have all the science at the wildlife center, all the observations and that kind of thing." He believed that all that needed to be done could be done at the center. However, Mr Roberts was disappointed that there were no tiger snakes to be seen on the days the students visited the center.

> It also would have been nice to get some credible results and things. ... with no sightings of snakes, that was pretty difficult. That could have been nice to actually see snakes, and observe their behavior. I always knew there was a challenge in that.

In spite of these differences between the members of the partnership, the project was successful because each member of the partnership worked toward achieving a positive outcome.

Integration of Values, Knowledge, and Skills in the School Curriculum

The state curriculum followed by Chelsea is underpinned by a set of values. Teachers were pleased to see how this integrated program enabled values to come to the forefront of students' activities. The program was able to demonstrate the values of social, civic, and environmental responsibility. This aspect was also high in Mr Roberts's planning of the program, who indicated that he looked specifically at environmental responsibility: "I use the endangered species that are here, and I talk about how the human impact has had an effect on endangered species."

Teachers also found that an advantage of working with projects with a strong community focus is the potential to integrate skills and abilities of other subject areas with the science content. For example, in describing how students analyzed the community survey results, Ms Edwards explained that, in addition to the science basis for the survey, there was the "math where they had to go over their percentages; they had to work out so much that we ended up doing quite a bit of math revision."

Ms Edwards noted other integrated curriculum skills incorporated into the students' learning, including "following directions, reading and comprehension skills, other English skills [such as] presentation skills, speaking clearly and slowly, and looking at your audience." She gave two further examples: computing, where the students learnt about graphing while learning how to use spreadsheet software, and graphics in art and design, where they learnt to use PowerPoint for their community night presentations.

Ms Mariott and Ms Jackson agreed with Ms Edwards's view on the integration of values and other subjects. Ms Mariott thought that the integrated project provided important opportunities to look at students' skills in a range of subject areas; however, from an assessment perspective, she remarked that she would have been very cautious. She thought there were too many constraints placed on students during the project for her to be comfortable to assess many of the integrated subject areas. She gave art as an example, explaining that she believed art was incidental to the project and, although it satisfied some of the content needed to be covered in art, the students' minds were on completing the activity, not doing their art work, and she thought assessment should be based on a specified task in art.

The overall successful outcome demonstrated that a great effort had been made by the students and teachers. Ms Edwards described it this way:

> Their level of commitment to excellence, their enthusiasm, their dedication, their personal development, and confidence was soaring. You had children there, I can think of one kid that had not done a public presentation before, huge personal growth for him, and it was an environment that enabled that to happen because everyone was pulling together, everyone was doing their very best and it gave us an opportunity to showcase the kids doing their very best and I think that was excellent, the actual learning on the subject area was enthusiastically embraced. I personally was happy with it as an outcome for the children.

Factors Affecting the Success of the LWTS Program

Planning, Communicating, and Goal Clarification

Prior to submitting the funding application for the project, Mr Roberts had sought cooperation from Chelsea's principal. They discussed the opportunity of integrating Mr Roberts's project into the school science program and mutually agreed that it was "a great idea." The principal approached Ms Edwards, outlining the project's merits. As the project application needed to be submitted very quickly, there was not much time to consider the implications carefully. Ms Edwards felt that "there wasn't a lot of consultation [with the staff involved] and I think that did cause some problems because we already had a full program."

Originally, Mr Roberts's plan was to do nearly everything at the wildlife center, but in reality much of the work was done at school. He explained that this occurred because

> there was a fair bit of pressure at the school … there were a lot of things happening and they had to slot this in with their program and so traveling out here doing sessions and then going back, it was fairly time consuming. So we felt that it was better value to stay at the school. The other sessions had to be

done here because it was very much outdoors, hands-on sessions looking at and observing the outside environment.

Ms Mariott found that for their students,

> The wildlife center wasn't an ideal learning environment. We were scrounging for space, for kids to have a table that was not cramped or crowded. They didn't have their own gear, they didn't have access to the material that they were meant to have, which was really integral in the end, to the end product, so I don't think it was an ideal resource base for us to work out of anyway.

Nevertheless, cooperation between the teachers and Mr Roberts was very good. On reflection, the teachers thought the project was very successful. Ms Edwards was the key person at the school and the one who liaised with the center manager; consequently, she spent considerable time and energy to make sure the project was successful. At the beginning, the class teachers were not clear about the project aims or its program. Ms Jackson believed that the project would not have been successful

> if it wasn't for Ms Edwards keeping it all together. It could have gone completely off the rails I feel. So next time I think everybody needs to get together to write the submission, to be very clear about what's going to happen and how it's going to happen, so that there are no big surprises at the end.

Use of the Community Resource

Ms Edwards, as the coordinator at the school, with Mr Roberts at the wildlife center were able to successfully involve students and the wider community in a program about environmental responsibility that could provide a long-term benefit to preserving the natural eco-system. The program provided a context where the school was guided by a partner from a resource in the local community.

This program would not have occurred without the funding provided. Although modest, it paid for the transport of children to the wildlife center, which the school otherwise could not have afforded. Ms Jackson explained:

> For a school like us, funding is really, really important because we don't run at a profit, we have low fees even compared with a lot of similar schools. So funds, for sure, are definitely really important for us.

From Mr Roberts's perspective, the funding was welcome: The center was subsidized—it has to operate on a fee-for-service basis—and the funding paid for supper on the community night. However, he considered it to be insufficient: "If we did something similar there would need to be more resources for the students and money to cover the administration because that was a lot of work." Nevertheless, he believed the project was a success. He noted that

> The connection between this highly relevant local community issue as a focus for promoting scientific literacy in the school and community was explicit throughout the project, with students enthusiastically taking ownership of a community problem that they were helping to solve by working scientifically.

Time Required to Complete the LWTS Program

The teachers at the school and the center manager both underestimated the time involved to produce a successful end result in terms of the community night. As a consequence, teachers put aside other work, disrupting their planned work schedule because they believed it was important to do so. Ms Jackson thought this could be rectified for future projects by being more involved in the whole planning process, as she believed it could be embedded in the curriculum, rather than an "add-on." Ms Mariott explained it this way:

> There was no way that whatever was achieved in those afternoon sessions at the center was anywhere near enough to get those tasks done. So even [though] we had done the [first] two weeks devoted to snakes, there was still afternoons that we had completely devoted to getting some tiger snake work done here at school.

Discussion

The choice of tiger snakes as the focus of this integration project lent itself to other learning areas of the curriculum as well as attracting the attention of participants. It also addressed, to the surprise of the teachers, other learning areas of the school curriculum, such as art, design technology, mathematics, and English. We think teachers were surprised because they had not realized that working with real-world issues is interdisciplinary and always requires more than one school subject to come to grips with the issues involved. Further, school–community projects promote a range of knowledge and skills, as shown by Kenney, Militana, and Donohue (2003), where a watershed program was developed to help teachers use their local environment to enhance classroom instruction. Other science educators have suggested how snakes can be used as a focus for learning in other areas, for example, Thomson (2003) has described a project in Kenya wherein the study of snakes assisted in providing links between languages and science, helping to preserve Indigenous education.

Despite some of the challenges involved, this research has shown that a school–community project can be a valuable initiative where schools and community resources can work together to explore relevant, real-life issues. Student motivation and focus was high, and teacher commitment exemplary. Nevertheless, before the commencement of such a project, careful consideration by teachers must be given to programming and planning, allowing the project to become part of the curriculum and not an addition, creating pressure on a crowded curriculum. Despite the time and effort involved, however, all three teachers and the center manager would be prepared to participate in a similar project in the future.

Focus Questions

1. What kinds of community resources are available in your local area? What steps would you need to take to be able to build in the use of at least one of these resources into your school curriculum?
2. If you were to manage a school–community project, how would you work to facilitate communication between all of the participants to ensure that the expected outcomes of the project were understood and were likely to be achieved?
3. Many school–community projects are based around environmental education. Others, like visits to an industrial site, are less obviously so. What kinds of knowledge and skills could students develop from their participation in such visits? What kinds of social, economic, and environmental values are able to be emphasized?

Suggestions for Further Reading

Powers, A. L. (2004). 'An evaluation of four place-based education programs'. *Journal of Environmental Education*, 35(4), 17–32.

In this article, Powers examines the characteristics of place-based education, which employs the resources, issues, and values of the local community, and fosters partnerships between schools and communities. The article provides useful schematics for successful ventures into environmental, place-based education. Powers's evaluations also suggest that these kinds of programs are beneficial for students with special needs, and encourage student motivation toward learning and engagement in school.

Rennie, L. J. (2011). 'Blurring the boundary between classroom and the community: Challenges for teachers' professional knowledge'. In D. Corrigan, J. Dillon, & R. Gunstone (Eds.), *The professional knowledge base of teaching* (pp. 13–29). Dordrecht, the Netherlands: Springer.

In this chapter Rennie shows how school–community projects are naturally integrated and indicates the range of knowledge, skills, and abilities associated with scientific literacy that can be developed through participation in these projects. She identifies the characteristics that enable them to be successful, and points out that, by their involvement in such projects, teachers often experience considerable professional learning.

References

Kenney, J. L., Militana, H. P., & Donohue, M. H. (2003). 'Helping teachers to use their school's backyard as an outdoor classroom: A report on the watershed learning center program'. *Journal of Environmental Education*, 35(1), 18–26.

Thomson, N. (2003). 'Science education researchers as orthographers: Documenting Keiyo (Kenya) knowledge, learning and narratives about snakes'. *International Journal of Science Education*, 25(1), 89–115.

11

FOCUS ON VALUES

Investigating Water Quality in a Local Lake at Kentish Middle School

Grady Venville

Introduction

School projects that are community based involve students learning important concepts and ideas, and completing tasks that are related to an important local community event, project, or development. For example, students might develop their knowledge of the local flora and fauna and geographical features to prepare a pamphlet that informs visitors and the public about a nearby park. An inevitable consequence of community-based school projects is that they do not fall neatly within the boundaries created by the disciplines, such as science, that form the principal structure of curriculum documents in most countries (Beane, 1995). This creates a kind of "curriculum tension" between the requirements to educate young people for civic responsibility, and, at the same time, educate them about the concepts and conventions of specialized subjects. This "curriculum tension" is most visible when teachers come to assess students' learning (Kelly, Luke, & Green, 2008).

This chapter explores a case study of a middle school involved in a local community project about the water quality of a local lake. It also examines the tension and difficulties that arose when teachers attempted to educate students about civic responsibility as well as the concepts and conventions of the disciplines such as science and social studies. The case story provides information about the way the project was implemented, the student learning that resulted from the project, and the approach taken to assessment. The discussion focuses on how the teachers were able to overcome the curriculum tension described above by using integrated approaches to assessment and reporting student achievement in a standards-based approach.

Kentish Middle School

Kentish Middle School is a purpose-built, government-funded middle school, and has a catchment of students predominantly from local, lower middle-class socio-economic suburbs. The case story presented in this chapter is about one learning community within Kentish Middle School of 120, grade 6 and 7 (10-, 11-, and 12-year-old) students. Five teachers worked with these students, including Ms Price, the learning community science coordinator, and Mr Keane, the learning community social studies coordinator. This case story was constructed using data collected from classroom observations, interviews with teachers and students, and a written survey of students. The physical structure of the classroom was open; the timetable was blocked but flexible, with time for collaborative teacher planning.

The learning community had become involved in a school, community, industry, partnerships in science project enabling it to receive a modest grant of $3000. The project had a 10-week implementation program and focused on the issue of water quality in a local lake. The project aimed to promote scientific literacy in the school and the community. In this context, scientific literacy was understood to mean using science content knowledge and the findings from the project to make decisions about the ecological health and wellbeing of the lake, and how those decisions involved social and civic responsibility. The project was explicitly intended to be integrated with the following statement incorporated into the initial proposal: "This integrated, real-world project will enable students to achieve outcomes from several learning areas from the state curriculum."

The state curriculum that Kentish Middle School follows is structured into eight subjects: arts; English; health and physical education; languages other than English; mathematics; science; social studies; and technology. Each subject, including science, is described at nine increasing levels of conceptual complexity that guide teachers' planning. The levels also were used, at the time of the project, for student assessment in a standards-referenced approach. Table 11.1 provides a summary of these conceptual levels in biology.

Another aspect of the state curriculum framework that is integral to this case story is that it is underpinned by a set of five core, shared values that teachers are required to integrate into their teaching and learning programs. The values include: 1) a pursuit of knowledge and a commitment to achievement of potential; 2) self acceptance and respect of self; 3) respect and concern for others and their rights; 4) social and civic responsibility; and 5) environmental responsibility. The educational context in the learning community in Kentish Middle School strongly reflected the core, shared values of the curriculum and the philosophies regarding the education of the adolescent child.

Case Story: Water Quality in the Lake—A Community-Based Project

The project topic, "Water Quality in Lake Wonthella," was selected by teachers at Kentish Middle School because the lake was in the students' everyday environment and could easily be accessed from their school and homes. Teachers felt that there was opportunity for students to make a difference to the quality of the water in the

TABLE 11.1 A Summarized Representation of the Levels of Conceptual Complexity for Biology in the State Curriculum

Level	The student:
Foundation	Demonstrates an awareness of own personal features and basic needs.
Level 1	Understands that people are examples of living things and that, like all living things, they change over time.
Level 2	Understands that needs, features and functions of living things are related and change over time.
Level 3	Understands that living things have features that form systems which determine their interaction with the environment.
Level 4	Understands that systems can interact and that such interactions can lead to change.
Level 5	Understands the models and concepts that are used to explain the processes that connect systems and lead to change.
Level 6	Understands the concepts and principles used to explain the effects of change on systems of living things.
Level 7	Understands the role of science in developing knowledge about systems and change.
Level 8	Understands how to assess the role of science in helping people to understand systems and change.

local lake. Mr Keane commented, "I like the idea of an 11-year-old being able to change the world, even just their own little bit of the world." All five teachers from the learning community worked collaboratively to plan and implement the project. Ms Price explained this process:

> In math, we did a lot of graphing, measurement, lots of measurement. ... So the way we worked, Ms Aldridge, who developed the math program, covered that in math. So we were already thinking at the beginning, when is the mathematical thinking coming in this topic? So when we needed to apply a math concept in science or social studies we asked her to do it in math. It's a bit like circular planning. The whole thing spirals on itself.

Each class within the learning community took a different project focus. For example, one class concentrated on the lake itself, another focused on houses and what people can do in their homes to contribute to a healthy lake, and another class looked at gardens and gardening approaches that are environmentally friendly. Mr Keane and Ms Price taught a range of interesting lessons to the group focusing on the lake. A variety of teaching techniques were used, including role-play (Lessons 9 and 16), games (Lesson 6), field excursions (Lessons 4, 5, 7, 8, 10, and 14), experimentation (Lessons 2 and 5), poster analysis (Lesson 3), modeling (Lessons 1 and 13), and project work (Lesson 15). The sequence of lessons was designed specifically to encapsulate problems associated with the lake.

Considerable time in the learning community schedule was allocated to the project. Large blocks were timetabled to enable classes to go on several excursions to the lake, and to other places, such as a water treatment plant. The content taught during observed lessons was strongly related to science and social studies. For example, the effects of pollution on human populations were considered in social studies (Lessons 11 and 13) while water testing for the effects of pollution was conducted in time allotted in the timetable for science (Lessons 3 and 5). Explicit connections were made between the activities by the teachers; however, the subjects were not identifiably discrete.

The teachers were able to draw from, and integrate the topic into, other learning areas including English, design technology, and art. For example, students were involved in a role-play and debate to explore land development and management issues around wetlands (Lesson 16). English-speaking skills were taught and assessed through this activity. Some students applied skills learnt in design technology to plan and build a model to demonstrate environmentally friendly city design (Lesson 15). In science, some students conducted experiments on the effect of dog excrement on water quality, and in art they designed and distributed pamphlets to the local community about the benefits of appropriate management and disposal of dog excrement.

Student Learning

Before the project, students thought about the lake from a personal perspective and how it was important to their families for recreation, including family outings like picnics and barbeques, and as an area where people could exercise and relax. After the project, when students were asked about what they learnt about the lake, they were able not only to talk about the nutrients that polluted the lake, but also to elaborate on the broader topic of human impact on the health of the lake with information about pollution, salinity, and habitat destruction. Moreover they were able to relate their personal experiences, the environmental consequences, and potential solutions. For example, prior to the project, Stacy said that her mother walks their dog at the lake every morning, and that her grandmother took her and her brother there when they were young to walk on the jetty and play on the grassed area. After the project, Stacy recalled learning about food webs through a role-play and about salinity with a game.

INTERVIEWER: Do you remember what animal you were in that role-play?
STACY: I was a snail.
INTERVIEWER: Right, and do you remember what you ate?
STACY: I ate algae.
INTERVIEWER: Right, and what ate you?
STACY: Um, I got eaten by one—a long-necked turtle. A turtle.
INTERVIEWER: Oh. Did you see any turtles in the lake?

STACY: Yeah, we saw a couple. Also we saw a bit of pollution—a witch's hat had been thrown in there. But we saw about two or three turtles. And lots of birds.
INTERVIEWER: Did you know much about food webs before?
STACY: Not really. I did learn a little bit about the oceans and food chains but this time it was about something that we went to visit and we could see the things in the food web down at the lake.

Later in the interview:

STACY: With the salinity, we learnt it in science; we learnt that how the salinity is like damaging the bushland and things like that.
INTERVIEWER: How did you learn about that?
STACY: We learnt about it basically with like, books and overheads and we did some experiments.
INTERVIEWER: Tell me about the experiments.
STACY: We had some soil and we had some salty water in it. And we used that. We also did a game for fun. We were like pretending to be trees and then we had to eat liquorice, but it was meant to be like salinity water. We were all trees and if we didn't like it we'd have to spit it out and then we'd die and we'd be replaced with new crops. And then the salt would sort of rise, and we learnt about it like that as well.

Tania had recently moved to the area—"We came here last year"—so she hadn't been to Lake Wonthella prior to the project. Tania had Ms Price for science and recalled conducting an experiment that helped her to understand the pollution in Lake Wonthella.

TANIA: We mostly did experiments, so like on pollution in Lake Wonthella.
INTERVIEWER: Do you want to tell me about those? Do you remember what you did for the experiments?
TANIA: Oh we had three different lake waters in different jars and then we got two tablets, and then we mixed them in and crushed them and however blue it goes, that's how much nutrients is in the water.
INTERVIEWER: Oh really?
TANIA: Yeah. Lake Natural, was—it hardly had any (nutrients) in it because it was natural and there was no drainage going into it, and then Lake Wonthella had a bit of nutrients in it because there was some drainage going into it. And then Lake Karebe had lots of nutrients because there was lots of drains going into it and feral animals and that kind of stuff around it.
INTERVIEWER: That polluted it?
TANIA: Yeah.
INTERVIEWER: So what do you think people living around Lake Wonthella should be doing?

TANIA: Not putting, like detergents and chemicals down the drains. Washing the car on the grass and not on the driveway so it doesn't run off into all the drains and that, all the chemicals.

Tania also recalled participating in an excursion to a water treatment plant and commented on people's waste water habits.

TANIA: Where they had waste water treatment plants and that. Yeah, that was—I like the waste water treatment plants because they had lots of like, you could see all the stuff that actually goes down the drain and how disgusting it is.

Approach to Assessment

Mr Keane explained in an interview after the project that the approach to assessment in this learning community is different from the traditional strategies of tests and examinations. The teachers collected data about the students' achievement in each of the learning areas in a more incidental but holistic way, so that, by the end of the teaching term, they had a number of pieces of evidence that allowed them to make judgments about each student's progress with regard to the biology progress maps (see Table 11.1) as well as progress maps in other learning areas and to provide feedback to parents. This approach was consistent with the state reform documents that specified, at the time of the project, that teachers were to use this standards-based form of assessment. Mr Keane explained further:

> We've never expected to have an assessment that actually gives everything in one time. We try and devise a lot of options for assessment, so the kids, if they are having a bad day, they don't get a bad mark. We don't say, "here's your biology test," which is conventional, "you got 28% or you got 78%," whatever it is … "That's your assessment piece for the year." We try to give it more of an incidental, I suppose, I don't think that's the correct word but, we will, as five teachers, we try and build in these holistic, these small assessment pieces and collect some data, collect other data, collect some more data. At the end of it, we can actually say something about the student.

Mr Keane explained how this method of assessment was used during the role-play activity that was conducted in Lesson 16. One role-play activity was used by the teachers for assessment in three learning areas including science, social studies, and English.

> The first is a role-play. Again, think of this as an assessment for English, science, and social studies. It's a role-play in a scenario where [there is] a lake, not unlike Lake Wonthella, and there's a development planned for the northern side of the lake. They have to role-play [stakeholders] in groups of five, one as developer, the Wonthella sailing club, the friends of the lake. … And we can draw out the application of ideas from a science idea. From the science idea we have to apply to a social situation, but using English language skills to do it.

Mr Keane acknowledged that the approach taken in the learning community to teaching, learning, and assessment might not be readily recognized as a conventional, disciplinary approach, and sometimes it was difficult to recognize and document the specific learning area outcomes that resulted from the project.

> But some days, we're not sure, you can walk in, are they doing science, are they doing social studies? No they're doing the lake. But look at the skills they're getting. Look at the outcomes they're producing. We're very focused on what happens at the end of the term rather than what happens at the end of the lesson. ... And our community is heavily focused on group tasks, working together, and kids having that respect for each other, and giving them that way of approaching things.

This assertion from Mr Keane is supported by the classroom observation of Lesson 15, where the students were involved in making group presentations on a concept of their own choice that they had learned during the term's work about the lake. This 1.5-hour session was used to assess the students specifically in English speaking and listening skills, but clearly demonstrated the students' learning in science as well as social studies. A vignette describing this lesson, Expo Day, is presented below, and includes the English speaking and listening assessment of the science-based presentations.

> This lesson was called "Expo Day" because all students in the community were involved in giving oral presentations about their work. The objective of this 1.5-hour session was that the students would make a group presentation based on the term's theme of "Saving the Wetlands." The students had prepared their presentations over the previous weeks and were told that it should be designed to educate other students on a concept learned during the unit. Students were assessed in English speaking and listening during this lesson as they had been learning the skills required to present to a group of students and the skills of being an attentive listener. They also had been exposed to a number of presentation techniques. The guiding criteria on the English assessment sheet included "matter," "manner," and "method," each of which had several subsidiary points.
>
> The whole learning community of 120 students was divided into five smaller "classes" within which five groups of four or five students made coordinated and planned group presentations to the other students. One teacher explained the expectations that you must listen if you expect others to listen to you and that you should watch the presentation with your eyes.
>
> At 11.45 a.m. the presentations began. The observer joined one group and watched the presentations from the five smaller groups. Summaries of these presentations follow.

Group 1 made a PowerPoint presentation called "Raindrop Man." The students read and role-played a water drop in the water cycle. They talked about going through the toilet and they clearly drew on some of the ideas they had learned on an excursion to the waste water plant (Lessons 7 and 8). The students had produced a booklet of the story with questions.

Group 2 role-played a television quiz show called *Catch a Natural Plant Quiz Show*. They had prepared a simple model of a television set from cardboard boxes; one student role-played the quiz host and others the contestants. The quiz included questions on the effect of heat on seeds, weeds in natural habitats, as well as water use.

Group 3 had built a model of the wetlands and prepared a script of their talk about the relationships between habitat and native animals. Their script described the ideal situation, and then they asked the audience questions, such as, "What would happen if you cut down the trees?" Each student spoke about an issue, for example, one student spoke about water as a precious resource in desert habitats.

Group 4 prepared a poster presentation about an "ecosmart" house and how houses can be designed to reduce greenhouse gases and save water. Each student contributed ideas; for example, Andrew talked about energy-efficient solar panels and the north/south orientation of the house, and Josh talked about including a rain water tank in the garden, planting drought-resistant plants, and including water-efficient appliances such as dishwashers, and low-volume sinks and shower heads.

Group 5 had prepared a large stick-on jigsaw puzzle with questions about the water cycle that the audience had to answer correctly to get the right, matching pieces of the puzzle. Questions varied, including about the safe storage of chemicals, how fertilizers impact on the wetlands, and how to use water wisely.

The civic and integrated nature of the outcomes from the project was clearly described by Mr Keane and is evident in the following quotation:

> There's real ownership of it. I can see that these kids are going to be the kids that can explain why you shouldn't feed bread to the ducks down at the lake, and they actually have the science behind them. They actually start to realize that there's the science behind all of water. So, [they might say] "Dad, no you can't pour oil down the drain." This is how a real product comes from my backyard to the lake. And from English they'll have the speaking and listening skills to actually deliver in that way.

Discussion

The evidence presented in the case story shows that, by being immersed in this integrated, community-based project, students not only learned scientific concepts, but also related these concepts to human action. Importantly, the students also developed the skills of communication and how to articulate their understandings and values with regard to the lake and associated environmental issues. The interviews with students, the survey, mind maps, and classroom observations clearly showed an increase in students' awareness of factors impacting on the health of the lake and public responsibility for the lake environment (Venville, Sheffield, Rennie, & Wallace, 2008). For example, the concept of the water cycle was used to help students understand environmental issues such as acid rain and the leaching of nutrients from garden fertilizers into the lake. The ideas expressed by students were clearly representative of the core-shared values underpinning the state curriculum.

Concepts and processes from several learning areas of the state curriculum, in particular science and social studies, were used to inform and understand aspects of the central issue of the wellbeing of the lake. For example, teachers developed a program that included the scientific process of testing the quality of the water of Lake Wonthella and made explicit links in social studies to human activities such as development and recreation.

The "curriculum tension" between educating for civic responsibility and educating for disciplinary outcomes is evident in this case story. The ways in which students demonstrated their learning were idiosyncratic and individualized. For example, the lesson *précis* in the vignette clearly shows that students in Group 1 focused their learning on the water cycle and demonstrated this by writing a story about a drop of water. The students in Group 3 focused their learning on habitat destruction and demonstrated this by building a model of the lake and the surrounding area. Moreover, the interview excerpts show that Stacy understood food webs in Lake Wonthella and the issue of salinity. Tania could explain the issue of nutrients in Lake Wonthella, how human activity impacted on the nutrients, and how the problem could be minimized through change of behavior. This idiosyncratic learning means that it would not have been possible to give the whole group of students a test on science content, or some other form of common assessment, for purposes such as comparability and reporting in grades.

The teachers in this case story addressed the issue of assessment by using the criteria, or levels, outlined in the state curriculum (see Table 11.1). This standards-based assessment strategy, used in many public schools in this state at the time of the project, gave the teachers the flexibility to document student learning by collecting various pieces of information, to collate that information within each of the learning areas, and then assess the students' learning against the generic criteria. This approach meant that a common assessment task was not necessary in order to give students grades. The standards-based assessment alleviated the curriculum tension because students were able to participate in the community-based project and to develop and participate in idiosyncratic science investigations with different science-based content. A similar case study is reported by Petrosino (2004), who found that the teacher of an integrated, project-based astronomy course differentiated between "testing" and "assessment," and "he made notes in his book about each student's progress on almost a daily basis. Thus, he provided students with detailed explanations of their progress" (p. 457).

While the curriculum tension was alleviated in this learning community, Mr Keane recognized the difficulties associated with the way the nature of the integrated, community-based project interacted with the subject-specific learning requirements of the curriculum. For example, he explained that the students need to understand the discipline structure of the curriculum before they move to a more traditional structure of a high school:

> So if they [the students] go into a conventional high school that's got [implications], you know from their first session, Thursdays, geography, you've got to do geography, whatever, in grade 11. We have to give them that structure to get into that. They can't just come out from our [school] saying, well, but we've been doing the lake.

We concur with Mr Keane, but take this reasoning a little further to suggest that, as there were no important examinations for these middle school students, the approach taken to curriculum and assessment was free from the constraints that face older, high school students, for example. In high school, when students often sit important standardized tests, there is generally a more clearly defined body of science knowledge that constrains what can be taught and learnt in order to enable the students to pass the tests and, in some situations, progress to college.

Conclusion

An integrated, community-based teaching and learning program can result in students learning concepts within the disciplines such as science, but also being able to apply these concepts in real-world contexts. The case story in this chapter showed how students learnt science concepts, like the water cycle, and related these concepts to human action such as the creation of acid rain. They also learnt how to express their ideas and values with regard to environmental issues related to the focus of the project, the local lake. The problem of assessing students' learning within the disciplinary structure of the state curriculum was addressed through the use of individualized, ongoing assessment, where students' progress was documented against a standards-based assessment scheme.

Focus Questions

1. How can standardized approaches to assessment, involving tests and examinations, support and/or create barriers to the implementation of integrated approaches to the teaching and learning of STEM subjects?

2. What forms of assessment are most likely to support the implementation of integrated approaches to the teaching and learning of STEM subjects? Why are these approaches likely to be supportive of curriculum integration?
3. How can assessment best be implemented and used to maximize the benefits of an integrated approach to curriculum?

Suggestions for Further Reading

Dawson, V., Lock, R., Brickhouse, N. W., & Crosthwaite, J. (2002). 'Teaching ethics'. In J. Wallace & W. Louden (Eds.), *Dilemmas of science teaching: Perspectives on problems of practice* (pp. 175–190). London: RoutledgeFalmer.

This book chapter considers the teaching of ethical issues in science and the ways that schools teach values. It raises very important questions for those of us who teach about ethics and the role of a teacher's own moral views and values in such teaching. The chapter includes a case story called "Playing God" by Vaille Dawson and commentaries from Roger Lock, Nancy Brickhouse, and Jan Crosthwaite.

Durán, R. P. (2008). 'Assessing English-language learners' achievement'. In G. J. Kelly, A. Luke, & J. Green (Eds.), *Review of Research in Education: What counts as knowledge in educational settings: Disciplinary knowledge, assessment, and curriculum* (Vol. 32, pp. 292–327). Thousand Oaks, CA: Sage.

This chapter in a special issue of *Review of Research in Education* considers assessment in more depth. In particular, it considers the issue of large-scale assessments that are intended to hold schools accountable for what students know and can do on the basis of their performance. In the context of English-language learners, the author considers alternative assessments that provide more valid information about learning capabilities and achievement.

References

Beane, J. A. (1995). 'Curriculum integration and the disciplines of knowledge'. *Phi Delta Kappan*, 76(8), 616–622.
Kelly, G. J., Luke, A., & Green, J. (Eds.). (2008). 'What counts as knowledge in educational settings: Disciplinary knowledge, assessment, and curriculum'. *Review of Research in Education: What counts as knowledge in educational settings: Disciplinary knowledge, assessment, and curriculum* (Vol. 32, pp. vii–x). Thousand Oaks, CA: Sage.
Petrosino, A. J. (2004). 'Integrating curriculum, instruction, and assessment in project-based instruction: A case study of an experienced teacher'. *Journal of Science Education and Technology*, 13(4), 447–460.
Venville, G., Sheffield, R., Rennie, L., & Wallace, J. (2008). 'The writing on the classroom wall: The effect of school context on learning in integrated, community-based science projects'. *Journal of Research in Science Teaching*, 45(8), 857–880.

12

REFLECTING ON CURRICULUM INTEGRATION

Seeking Balance and Connection Through a Worldly Perspective

Léonie Rennie, Grady Venville, and John Wallace

Introduction

The ten case stories described in the previous chapters covered a variety of ways in which teachers integrated part of their science, mathematics, or technology curriculum, either between these subjects or with other subjects. Although none of our schools included the subject of engineering in its curriculum, many chose an engineering project as an outcome of the integrated work. Each story of integration was different. The Greenwich Public School ice hockey unit (Chapter 3) was designed and implemented by a single teacher, whereas the entire curriculum at Seaview Community School (Chapter 5) was organized around a literacy focus, something that all four teachers in this small school supported and implemented in their own classrooms. All of the teachers believed that their efforts were beneficial for students, and that there was considerable value in linking their in-school activities with events outside of school.

What can we learn from these case stories? In particular, what can we learn in answer to the two questions we posed in Chapter 1: Do students need strong disciplinary knowledge to cope in today's changing world, or do they need cross-disciplinary, integrated knowledge? Should students learn about local issues, or focus on issues of global significance?

In this chapter, we revisit the ten case stories and draw from them the lessons we learnt about the kinds of conditions and structures that facilitated teachers' efforts in implementing an integrated curriculum in STEM subjects, and the challenges that needed to be overcome to do this successfully. Based on this discussion we suggest that a more holistic view is needed to consider the kinds of curricula that will be of advantage to students as they leave school and begin their life after school. We call this holistic view a Worldly Perspective, and we describe how it can help frame STEM curricula.

Implementing Integrated Curricula: What Helps and What Doesn't?

A summary of the ten case stories is provided in Table 12.1. It lists the schools in chapter order, identifies the teachers involved, and indicates highlights for each school according to the conditions that supported integration, the challenges teachers experienced, and the key findings that might be transferable from each case story to other schools. From this summary, and in the context of the outcomes from our broader research program (Rennie, Venville, & Wallace, 2012; Wallace, Sheffield, Rennie, & Venville, 2007), we identify six attributes that impact on the success of an integrated program. These attributes are small and stable learning environment, leadership, teacher team activities linked to the classroom, in-school planning time, flexible timetable, and community links. In the following sections, we describe each attribute and illustrate its discussion with examples from our case stories.

Small and Stable Learning Environment

An essential attribute of successful integrated programs was a small and stable learning environment. In two schools, Greenwich Public School and Southern High School, the program was implemented by one teacher with her class. Both were project-based examples of integration, focused on building an ice hockey rink and a bridge, respectively, and in each case the teacher had used this unit with previous classes. Programs in other schools (Beachville, Leaside, and Rinkview) were implemented by two teachers who had previously worked together. The integrated program at Kentish Middle School involved five classes with five teachers working collaboratively. In this large school, students and their teachers were divided into learning communities, a school sub-community in which an interdisciplinary team of teachers shared responsibility for their group of students. An important aspect was that the programs, staff, and relationships within these teams were well established and the learning environment was stable. Stability was also a feature of the small, independent Chelsea Elementary School, with only two classes of mixed grade 4 to 7 and their two teachers who frequently combined their classes. Seaview Community School was also small; its four teachers were geographically isolated from other schools, they lived on the school grounds, and formed a very stable learning community.

At Gosport Community School, the grade 8 classes involved in the integrated program were two of six taught by a group of teachers who formed a learning team within a much larger school structure. While the teachers in the learning team shared an office, and thus the same physical environment, they tended not to share ideas about teaching, and, unlike the learning team at Kentish, there was no coherent learning environment. As a result, science teacher Ms Potter found herself isolated in her attempts to implement a program on the theme of access for disabled

TABLE 12.1 Summary of Case Story Approaches, Conditions, Challenges, and Outcomes

Chapter, School and Teachers	Integration Approach	Supporting Conditions	Challenges Experienced	Key Outcomes
Chapter 2 Leaside High School Mr Batani Ms Davis	*Project-based Approach:* A mathematics and science teacher and a technology teacher both taught a grade 8 class of mixed ability students. During each of the 10-week terms, the students built projects that combined knowledge of these three subjects.	• teaching teams • collective vision • dedicated teaching space • meeting time • flexible timetable	• teachers had not identified expected outcomes and this impeded assessment of the learning from the project • loss of students' interest when frustrated by failure	• When outcomes were clearly stated at the beginning of the project, both formative and summative assessment were improved and students were motivated to reflect thoughtfully on their progress. • Success and motivation are mutually enforcing.
Chapter 3 Greenwich Public School Ms Perry	*Project-based Approach:* One teacher and her 6th grade class worked on a 5-week integrated unit on the theme of ice hockey. Students were expected to select players for a fictional team, develop a game schedule, and build an ice hockey rink to scale.	• teacher fully cognizant with requirements for integrated subjects • dedicated teaching space • flexible timetable	• teacher workload	• Students were unable to transfer knowledge from one context to another unless connection was pointed out. • The integrated context enabled the teacher to identify and remedy students' misconceptions.

TABLE 12.1 (continued)

Chapter, School and Teachers	Integration Approach	Supporting Conditions	Challenges Experienced	Key Outcomes
Chapter 4 Southern High School Ms O'Reilly	*Project-based Approach:* The grade 9 technology class was set a bridge-building project which incorporated knowledge of science, mathematics, engineering, design, and construction. The aesthetics of the bridge were judged by the English teacher.	• teacher with strong knowledge of engineering • flexibility of unit content • appropriate teaching space	• some students didn't recognize the utility of their subject-based knowledge in science and mathematics to the engineering task	• Inclusion of time and budget issues enabled connection between local and global goals.
Chapter 5 Seaview Community School Mr Lanyon	*Cross-curricular Approach:* A school-wide literary focus aimed at assisting students to learn English, which underpinned all subject areas and was supported by cross-curricular initiatives, such as horticulture relating to the school garden.	• administrative leadership • flexible structures and time-table • small school size • community support for school's objective to help students learn good English	• transient student population prevented longer-term projects related to subject curriculum content	• Continuing long-term cross-curricular projects such as horticulture can be sufficiently flexible for students to participate even with transient attendance. • Involvement of community in school objectives ensured support on language focus.

TABLE 12.1 (continued)

Chapter, School and Teachers	Integration Approach	Supporting Conditions	Challenges Experienced	Key Outcomes
Chapter 6 Beachville High School Ms Wade Mr Norris	*Synchronized Approach:* For a grade 9 applied class, the geography and science teachers collaborated to teach the 5-week electricity and energy use unit separately but in parallel. In geography and in science, cross-curricular links and concepts were reinforced.	• planning time prior to implementation • committed teachers with collective vision • continuous communication with research team	• teacher workload • lack of teacher planning time during implementation	• Students were able to recall and apply in one subject concepts that had been taught in the other.
Chapter 7 Rinkview Public School Ms Charlie Ms Jane	*Project-based Approach:* Two grade 8 teachers joined classes to form a pod of 50 students to jointly teach a 6-week project-based unit on "Making and Marketing a Toy." The students surveyed peers about possible toys, and designed, built, and marketed their toy to other students.	• teaching teams • collective vision • dedicated teaching space • meeting time • flexible timetable	• teacher workload • limited teacher content knowledge • multiple goal expectations	• The collaborative approach helped teachers learn from one another, and modeled cooperative learning for their students. • Students were well motivated with a high degree of on-task behavior. • Multiple and competing goals meant that the learning focus was sometimes confused.

TABLE 12.1 (continued)

Chapter, School and Teachers	Integration Approach	Supporting Conditions	Challenges Experienced	Key Outcomes
Chapter 8 Gosport Community School Ms Potter Mr Birchwood	*Thematic Approach:* Two grade 8 classes engaged in a 10-week module about community access for disabled people. Disability was considered in science, social studies, and physical education.	• school support for the project • supportive lead teacher	• teacher inexperience • lack of support from other staff • teaching in isolation from other subject teachers • disaffected students	• A community focus does not necessarily engender enthusiasm in students; effective selection and implementation of learning activities is necessary. • Integration is facilitated by confident, experienced teachers.
Chapter 9 Mossburn School Ms Rahim Ms Haslam Mr Caspian	*Project-based Approach:* The grade 8 academic extension class participated in a 10-week-long project related to the design and construction of a model house. Teachers taught their own subject area but included aspects related to modern living.	• administrative support • leader experienced in the project • initial planning time	• leadership was not shared • some teachers could not see relevance to their own subject • crowded curriculum • common assessment across year level meant that the project was an additional imposition on time	• Without shared leadership, teacher commitment, and sufficient time, integration is unlikely to succeed.

TABLE 12.1 (continued)

Chapter, School and Teachers	Integration Approach	Supporting Conditions	Challenges Experienced	Key Outcomes
Chapter 10 Chelsea Elementary School Ms Edwards Ms Jackson Ms Mariott	*Community-focused Approach:* Two combined grade 4 to 7 classes worked closely with a local wildlife center on an integrated program aimed at understanding, and living with, tiger snakes. All school subjects were involved and the 7-week-long focus culminated in a community presentation by students.	• lead teacher with flexible time liaised between the center manager and teachers • funds to transport children to the center, and pay for their staff time • dedicated teachers who allocated additional time to ensure successful outcomes • strong subject matter knowledge of the center manager	• mismatch of expectations between teachers and center manager, relating to time and the nature of the activities involved	• The focus on the culminating event (the community night) motivated all participants to ensure success. • Focus on a real community issue facilitated integration of subjects, enabled a strong underpinning of values, and demonstrated connectedness between school work and the community.
Chapter 11 Kentish Middle School Mr Keane Ms Price	*Community-focused Approach:* A learning community of 120 grade 6 and 7 students undertook a 10-week, extensive study of the local wetland with an integrated approach involving water quality, land use, ecological studies, recycling, and economic factors.	• collective vision • regular planning time • flexible timetable • proximity of the wetlands to the school • assessment program accommodated diverse student outcomes	• teaching out-of-field by some teachers • surface learning in some conceptual areas	• Recognition of importance of the health of the lake provided focus for cognitive and affective learning, empowering students. • A student-centered approach enabled students to follow own topics of interest, promoting independence and self-direction.

people, and there was little integration with other subjects, apart from some shared activities with social studies. At Mossburn School, the integration focus was the construction of a model house in the grade 8 academic extension class. However, the teachers of this class were mostly concerned with teaching their own subjects to ensure that their students passed the common test, and made little effort to include aspects related to the house. Although there was some synchronous teaching with English and social studies, the actual model house construction project was left to the leader of the project and was not completed by the end of the term.

Leadership

Leadership was important to the success and sustainability of integrated programs; it could reside in the role of the school principal, a team leader, or shared between teachers. At Seaview Community School, the principal, Mr Lanyon, was the natural leader of his three teachers. He had a hands-on, transformative style of leadership, but responsibility for integrating literacy throughout their curriculum was shared with the class teachers. In some schools, like Kentish Middle School, the principal was supportive in providing encouragement, but also resources, like meeting time, and flexibility in curriculum.

Other kinds of leadership resided in the teachers themselves. At the two schools where integration involved a single teacher, as at Greenwich and Southern, these teachers were their own "leaders." At three schools, Beachville, Leaside, and Rinkview, two teachers shared leadership, taking equal roles in planning and implementing the project. At Chelsea Elementary School, the lead teacher, Ms Edwards, was identified as the program leader. The two class teachers, Ms Jackson and Ms Mariott, were committed elementary teachers but professed to know little about science. Ms Edwards was very interested in environmental science, she was well respected for her organizational skills and knowledge, and the other teachers happily followed her lead. Ms Edwards was the person who liaised with Mr Roberts, the manager of the wildlife center, and translated his ideas into action at school. This liaison was crucial for the success of the project.

The most pervasive integrated program was at Kentish Middle School. Ms Price and Mr Keane were the learning community leaders for science and social studies, respectively. Jointly, they led the integrated program; however, the other three teachers in the learning community were enthusiastic, committed to the program, and occasionally took the lead. This was the kind of leadership Rennie et al. (2012) found in the most successful integrated programs: leadership distributed across teaching teams, commonly referred to as teacher commitment or enthusiasm, with teachers sharing responsibility and contributing jointly to the program.

The least successful programs, in terms of achieving their objectives, were at Gosport and Mossburn schools. At Gosport, Ms Potter was concerned about an absence of effective leadership. While all teachers in the learning team initially agreed to a joint plan for the "Making a Difference" project, most of them retained their discipline-focused plans for teaching and did not contribute to, or share responsibility for, the project in any meaningful way. Mr Birchwood, the team leader, was supportive of the project and also of Ms Potter, who he could see was struggling to cope with her difficult classes. In this case, the program had a leader, but as an inexperienced teacher Ms Potter was unable to share the leadership.

Mossburn School had a dedicated and experienced leader for the model house construction project in Ms Rahim, because she had taught this integrated project in previous years with Ms Haslam. Ms Rahim did not teach the grade 8 class in the year of our research, but she endeavored to organize the project with the other subject teachers, planning how they might synchronize related topics with the model house project. However, the house project was identified as "Ms Rahim's project," and apart from Mr Hashim and Ms Haslam, who remained committed to the construction project and linked their social studies and religious education lessons to the house theme where possible, Ms Rahim had little support. The students' construction of the model house occurred during the science classes, supervised by Ms Rahim, with the science teacher, Mr Caspian, generally uninvolved. In the case of both Gosport and Mossburn, the less than hoped-for outcomes from their projects may be attributed to the difficulty of maintaining teachers' active involvement. This demonstrates the close correspondence between shared leadership and a small and stable learning environment.

Team Activities Linked to the Classroom

In successful programs where more than one teacher was involved, teacher team activities were closely connected to classroom instruction and included such things as shared curriculum materials, team teaching, collaborative development of themes, and coordination of assessments. At Beachville for example, Ms Wade and Mr Norris, the science and geography teachers, ensured that the topics in their respective energy units were taught synchronously. In schools like Kentish, where there was shared leadership and teachers had the same grade-level classes, these kinds of cooperative and collegial activities were the norm. For their in-school work at Chelsea, Ms Jackson and Ms Mariott combined their classes, and Ms Edwards worked alongside them. The students at Chelsea participated in all of their project activities as a single group; they visited the wildlife center together with their three teachers while Mr Roberts, the center manager, directed activities. Here the team activities exemplified the small and stable learning environment that existed for teachers and students at Chelsea.

At Gosport, some shared activities between Ms Potter and Mr Birchwood, the team leader, supported the integrated project. Due to their lack of interest in the project, however, there was no evidence of linked activities among the remainder of the learning team, who were focused on teaching their own subject. Further, Ms Potter's science class was taught in a small, specialist science laboratory separate from the regular open-plan team classroom, and this was a disincentive for sharing activities or team teaching. There was also limited sharing of ideas and activities at Mossburn School. The opportunities for Ms Rahim and Mr Caspian to team teach during the construction of the model house in science classes were not taken up. Further, Ms Rahim's efforts to coordinate the assessment activities related to the model house project were largely unrewarded because teachers went ahead with their own subject assessment plans.

In-School Planning Time

Quality planning time was a very significant enabler of the integrated programs, particularly when more than one teacher was involved. It was closely related to team activities that had to be planned and carefully organized to bring to fruition. Lack of time was a common lament among the teachers with whom we worked. In our broader research program, lack of planning time accompanied by a heavy workload led to the demise of some integrated programs, hampered others, and worked against the building of collegial relationships (Rennie et al., 2012). The combined pressures to "get through" the course content and prepare students for common tests made it difficult for teachers to cooperate on any degree of integration unless in-school planning time was specifically scheduled. Time formally set aside for teachers to collaborate facilitated innovation; it also served to indicate that such work was important and valued by the school administration. However, there needed to be leadership and commitment to use that time for planning, instruction, and assessment strategies for innovation to work. At Gosport, the middle school team had one hour of planning time each week structured into the timetable, but the meetings were often concerned with logistical rather than teaching and learning issues to do with the integrated project. At Mossburn, Ms Rahim held two initial meetings about the model house construction project, but lack of further common time reduced the likelihood of collaboration.

Some small integrated programs were successful even when there was no scheduled planning time. For example, having planned their synchronous teaching relating to their energy units at Beachville, Ms Wade and Mr Norris had little time to compare progress during the implementation. There was no scheduled planning time at Chelsea Elementary School, but Ms Edwards with her flexible hours was able to ensure that all teachers knew what was happening, particularly in terms of the out-of-school activities with Mr Roberts at the wildlife center.

Flexible Timetable

The scope for integrated, student-centered programs was greatly enhanced when the central school timetable established only the broad structure of the curriculum and allowed pedagogical decisions about student grouping, teaching time, and space allocation to be devolved to the teaching team. These flexible timetables had a common goal of lengthening blocks of learning time and reducing the number of transitions students needed to make between subjects, teachers, and other activities. When the program required students to work off-site, such as the community wetlands-based programs at Kentish and Chelsea, teachers had sufficient time to transport students and complete lengthy activities. Timetable flexibility also allowed for community experts to visit the school at times convenient to the visitors. Gosport was able to use some time for this, and to take students off-site to the basketball stadium.

Flexible timetables allowed longer time slots to make progress on engineering projects. At Mossburn, the fixed timetable was a limiting factor in building the model house because construction time was reduced by the need to set up equipment outside the classroom and pack it away again. At Leaside, Rinkview, and Southern, a dedicated, appropriate workshop space for constructing the various projects facilitated the completion of the project-based programs.

It was always easier to vary the timetable to accommodate teachers' plans for integrated programs in elementary schools and, to some extent, middle schools, compared with senior schools. Kentish and Rinkview did this successfully, with the learning team teachers regularly meeting and planning their activities. Small schools, such as Chelsea, had considerable flexibility because communication between the two class teachers and the lead teacher facilitated arrangements to vary the length of sessions or to go off-site. Seaview was also small, and, although it was a K-10 school, the three class teachers had non-overlapping grade cohorts, and could vary their time as they wished. This assisted the horticultural program because students could be available to do gardening activities as dictated by the season.

Community Links

The sixth and final attribute of the integrated programs described in our case stories concerned linking the curriculum content and classroom activities to the external community. This attribute highlighted the importance of bringing understanding of policy and practice regarding an integrated curriculum to the community, and that the projects were relevant to the local community and involved collaboration with community members. We identified two levels of community links: information and action. At the simplest information level, the students at Greenwich sought information from community sources to obtain the dimensions of ice hockey rinks in order to build their model rink to scale. In other schools, there were occasions when community members came into the school to advise on particular aspects of the integrated program. For example, both Gosport and Kentish had community experts visit the school to talk to students. The community night at the wildlife center was an occasion when Chelsea students displayed their work to parents and others in order to exchange information about the tiger snakes that were an integral part of the lake ecosystem.

The action level of linking with the community involved a greater level of involvement with, or contribution to, the community. The programs at Kentish and Chelsea involved partnerships with community institutions. Students at Kentish and Chelsea visited their local wetlands and interacted with community personnel during their visits. Their activities involved more than the seeking or exchange of information; they involved interaction between partners.

The "Making a Difference" project at Gosport Community School was instigated by a person external to the school who, on behalf of the local disabled community, wished to inform and engage students in understanding the importance of access. This was an authentic project that had strong visual and practical links with the local community, and the students enjoyed the out-of-school activities, such as playing basketball in wheelchairs. Unfortunately, in-school issues limited the potential benefits of these community interactions to students' understanding of the related science and technology. Compounding the issues discussed above were Ms Potter's inexperience and lack of well-developed pedagogical content knowledge. With Mr Birchwood's support, she struggled through, but her experiences in the unit underline the importance of teacher knowledge and experience in offering an integrated curriculum, and emphasize the importance of mentoring for new teachers.

In this part of the chapter, we have identified the attributes contributing to the success or otherwise of implementing an integrated program. It is clear that, although each attribute has importance, it is not necessary for all to be present; rather it is the combination of attributes and the dedication of the teachers that are important determinants of successful outcomes. The teachers in our case story schools believed that their integration attempts were beneficial for the students, and some also thought that an integrated approach was an effective way to learn about things outside of school. However, decisions about the implementation of integrated curricula were guided by what could be fitted into their school context.

The Culture of Schooling and the Nature of Curricula

It is clear that integrated approaches to curriculum are implemented with varying degrees of commitment and with varying life-spans (Wallace, Rennie, Malone, & Venville, 2001). Further, success is often idiosyncratic, dependent on the motivations and aims of the participants and mediated by their educational context. As illustrated by the case stories in this book, curriculum integration challenges what Tylack and Tobin (1994) called the "grammar of schooling," that is, the culture present in schools, as evidenced in the customs, ceremonies, and artifacts of everyday school life, including the tradition of organizing the curriculum around subjects. As many teachers in our case stories found, the traditional culture of schools is very difficult to change, particularly in secondary schools where the ethos of teaching and learning tends to revolve around subject departments.

The culture of schooling is reflected in two kinds of contextual factors: school traditions and administrative policies, such as timetable organization and staffing arrangements; and teacher variables, such as pedagogical content knowledge, beliefs, and instructional practices. Pang and Good (2000) found these factors to be important determinants of the success or otherwise of integrated programs. Our research findings were very similar; they demonstrated strong relationships between educational context and the way that integrated programs were implemented. Five of the six attributes discussed above were contextual factors. The sixth factor, community links, can also impact school culture through parental expectations, which often push to maintain the status quo within the school curriculum (Wallace et al., 2007). Parents' expectation is often that schooling should be academically oriented, emphasizing individual, written work and focused on examinable science concepts and ideas (Kaplan, 1997), something that seems to be at odds with an integrated curriculum that breaches the boundaries of traditional school subjects.

We found that implementing and sustaining an integrated approach to the curriculum is challenging for teachers, administrators, and others concerned with curriculum reform. The challenges derived from a curriculum approach, which by its nature is flexible, multidisciplinary, and democratic, colliding with a schooling context that is rigid, disciplinary, and hierarchical. Changing the status quo, or context of schooling—what Hall and Kidman (2004) referred to as the "strategic direction, policies, conditions and many of the rules that govern the way that teaching, learning and assessment take place" (p. 334)—appears to be a key factor in the introduction of integrated practices. Given this, it is not surprising that the efforts to integrate we observed over time were differentially successful and difficult to sustain.

In the next sections, and with a broader educational perspective in mind, we examine more closely three issues that seem to capture the essence of the tension that exists between discipline-based and integrated curricula. These are measuring learning, the purpose of STEM curricula, and knowledge and power.

Measuring Learning

Kelly, Luke, and Green (2008) pointed out that many current educational debates include an underlying assumption that there is an unchallengeable corpus of canonical, disciplinary received wisdom. The result is curriculum documents that espouse key criteria, standards, or educational outcomes narrowly focused on what is readily measurable or amenable to achievement testing. Many school systems have a high-stakes testing regime that determines students' future educational and potential career paths. This kind of external assessment has a strong influence on what is taught. Au (2007) demonstrated that the primary effect of high-stakes testing is to narrow the curricular content, fragmenting subject area knowledge into easily tested pieces, and encouraging a more teacher-centered pedagogy. This cycle of discipline-based curriculum, teacher-centered pedagogy, and assessment of content knowledge, works against the inclusion of other kinds of learning activities into the curriculum.

Schools that focus on preparing students for high-stakes tests limit their opportunities to engage in contextual, issues-based learning, particularly if teachers are not skilled in assessing these kinds of learning. Integrated curricula often require teachers to teach out-of-discipline, impacting both their confidence and ability to teach and assess (Kruse & Roehrig, 2005). It is easy for teachers in high schools to use common, content-based summative tests that, because they are used across classes, are assumed to be reliable and fair. What these tests do not do, however, is measure the impact that curriculum integration may have on attendance, student discipline, knowledge of academic resources, study habits, student enthusiasm, and student engagement, for which there is mainly anecdotal evidence (Hurley, 2001). Some studies have incorporated broader and more holistic perspectives into their evaluation of student learning, focusing on outcomes such as student motivation, attitude, cooperation, and capacity to transfer and apply knowledge. For example, Ross and Hogaboam-Gray (1998) found that students studying an integrated science, mathematics, and technology course improved their ability to apply shared learning outcomes, motivation, ability to work together, and attitudes to appraisal of group work.

It is important to understand and accept that integrated approaches necessarily incorporate both disciplinary and interdisciplinary objectives, and that the outcomes cannot be assessed authentically with a simple achievement test. Secondary analysis of one of our case studies (Rennie, Venville, & Wallace, 2011) demonstrated that outcomes needed to be measured from more than one theoretical perspective to appreciate the full extent of students' learning when they were involved in integrated project-based tasks. Mr Keane explained how the teachers at Kentish Middle School used continuous assessment to provide solid evidence of student performance. The positive impact on learning of this kind of formative assessment is well established (Black, 1998). We saw this also at Greenwich Public School; when Ms Perry found that changing the context of problems on proportion from the mathematics class to the model ice rink revealed gaps in students' understanding, she was able to remedy them. Teachers are in the best position to provide dependable assessment of students' performance because they are there, with students, in the classroom. However, as we discovered at Leaside High School, unless teachers could clearly articulate the project-based outcomes of their program, they found it difficult to assess students' achievement of them. More attention must be given to finding ways to measure a broad range of learning outcomes and allow more effective assessment of the nature of student learning from integrated curricula.

The Purpose of STEM Curricula

Increasingly, the purpose of learning science (as well as technology, engineering, and mathematics) in school has been described in terms of the need for a scientifically and technologically literate population, that is, to provide an education that will be of value to students over a lifetime, irrespective of their careers (Osborne & Dillon, 2008). Traditionally, high school science (and mathematics) has been seen as a passage to higher education and career choices for talented students interested in science (Aikenhead, 2006). However, the large majority of students who do not wish for a career in the STEM fields would benefit from a curriculum that prepares them to deal with decisions in their after-school lives. For example, the US National Research Council's Board on Science Education's vision, stated in their K-12 Framework, has two major goals for science and engineering education: "educating all students in science and engineering and providing the foundational knowledge for those who will become the scientists, engineers, technologists, and technicians of the future" (National Research Council, 2011, pp. 1–2). This is a considerable challenge that must take into account the interdisciplinary nature and complexity of STEM as it is practiced in the world today, together with the social and political pressures that shape it. Jenkins (1999) referred to "transdisciplinarity" as a way to address these pressures. He argued that "this means constructing … curricula that enable young people to engage [reflexively] with science-related issues that are likely to be of interest and concern to them" (p. 707).

A curriculum that is "of interest and concern" to students is particularly important for those students who are not the elite and for whom an academic, disciplinary-based approach to STEM has little relevance. The use of science-related contexts as a means of introducing science concepts is one approach (e.g., Gilbert, Bulte, & Pilot, 2011) that holds promise of leading students to a deeper understanding of concepts (Rivet & Krajcik, 2008), and an awareness and appreciation of the use of science and technology in society (Stocklmayer, Rennie, & Gilbert, 2010). Engaging students in STEM situated in social and cultural contexts requires breaching the boundaries between the various subject areas and introducing a more integrated, interdisciplinary approach to the curriculum. The Board on Science Education (National Research Council, 2011) refers to cross-cutting concepts "that bridge disciplinary boundaries, having explanatory value throughout much of science and engineering" (pp. 1–2). These concepts have also been called "unifying themes" or "big ideas," but what they emphasize is the common structure of these disciplines and the opportunities that exist for using them in interdisciplinary ways.

Knowledge and Power

We found many teachers, parents, and even students who considered school subjects with everyday, integrated knowledge to be "soft," that is, not easily tested, subjective, and open to debate. Subjects containing "hard" academic knowledge are those that are testable, objective, and well established (de Brabander, 2000). It seems that the more discipline based the subject, the higher its academic status and, conversely, the more integrated the subject, the lower its academic status. This perception bestows considerable power on discipline-based subjects. Young (2008) defined powerful knowledge with reference to what knowledge can do or what intellectual power it gives to those who have access to it. He claimed that the main reason parents send their children to school is the hope that they will gain powerful knowledge and that powerful knowledge is gained through studying the disciplines.

The case story at Kentish Middle School provided examples of integrated instruction that contrasted with the view that only highly disciplinary knowledge can be considered powerful knowledge. In an integrated, community-based science project underpinned by values relating to social and environmental responsibility, there was evidence that the students learned important science concepts. Students' learning in science was idiosyncratic and integrated with concepts from other subjects, but students could relate these concepts to issues in their local environment and engage in critique and debate about them. The interdisciplinary approach provided the students not only with powerful scientific knowledge but also with powerful values in civic responsibility, power to think about the problems and issues relevant to them, power to communicate and debate these issues, and power to evaluate ways that these problems and issues can be addressed. We saw a similar outcome at Chelsea Elementary School's community night, where the children's expertise about tiger snakes and how to cope with them exceeded that of most of their parents and other community members.

We reiterate: curriculum integration does not mean that disciplinary knowledge is ignored. Dealing with integrated issues and problems in science and mathematics, for example, requires the application of the cognitive and practical tools of the disciplines, including subject matter knowledge and rigorous explanations, as well as a range of cross-disciplinary knowledge and skills, such as problem solving, creativity, and argumentation. When students engage in this kind of curriculum, they develop a much broader range of skills and understanding than content knowledge alone. However, we need better assessment practices that can measure authentically what students have learned and how useful or powerful those learning outcomes can be.

A Worldly Perspective

We have established that, although disciplinary approaches dominate curricula world-wide, the real world of contemporary science, technology, engineering, and mathematics is multidisciplinary. Based on our own research, illustrated by the case stories included in this book, and supported by the research literature, we pointed out that integrated curricula challenge the status quo of schooling and traditional school subjects, such as science and mathematics. We demonstrated how assessment practice is dominated by a focus on measuring the learning of content knowledge, and that learning from integrated curricula is more difficult to measure and consequently often underestimated. We drew attention to the increasing emphasis on scientific, mathematical, and technological literacy as a central purpose of education.

We showed how integrated, contextualized approaches to the teaching and learning of STEM subjects can develop cross-disciplinary literacy, by providing students with opportunities to solve practical, real-world problems and participate in debate and decision making. Finally, we challenged the notion that powerful knowledge comes only from studying tightly defined school disciplines by showing how integrated teaching and learning can offer students powerful knowledge of a different kind by giving them the knowledge and skills to think critically about issues in the world around them.

Although we contrasted disciplinary and integrated approaches to curriculum, we regard these opposing positions as complementary, not incompatible. Our case stories revealed many ways that curricula were integrated, but all demonstrated some combination of both disciplinary and integrated knowledge. We found that integrated approaches included relevant disciplinary knowledge and constructs, but these were applied, and therefore learned, in different ways. Further, we found that integrated approaches that deal with local issues and concerns provided students with broad-based knowledge and skills that could be transferred to larger, global issues and problems. We believe that a more holistic perspective is needed to characterize a curriculum that encourages students to develop scientific, mathematical, and technological literacy. From this holistic perspective, knowledge of the disciplines is balanced with knowledge of integrated issues and how they are dealt with outside of school. Further, this holistic perspective includes the notion of connectedness between local contexts and issues and more global concerns.

We use the term Worldly Perspective (Venville, Wallace, Rennie, & Malone, 2002) to describe a holistic view of knowledge grounded in students' experiences, relationships, and contexts. The separation of integrated and disciplinary knowledge does not describe the world that students experience outside of school. From a Worldly Perspective, integrated and disciplinary forms of knowledge can be considered together: overlapping rather than mutually exclusive. A Worldly Perspective acknowledges high-status subjects such as physics, chemistry, and calculus, but allows these subjects to evolve within a broader framework than currently exists in most school contexts. This broader framework means that worldly knowledge is connected in some way to the experiences, contexts, and needs of the students' school community, and that these local contexts need to be well connected with global communities, global ways of thinking, and global ecologies.

In summary, we think of the Worldly Perspective as having two dimensions that can be used to evaluate the power of knowledge and learning that is likely to be developed through different approaches to STEM curriculum. The first is a *knowledge* dimension that can be used to indicate the degree of balance between disciplinary knowledge and integrated knowledge. The greater the balance between these types of knowledge, the more the curriculum is likely to provide students with powerful knowledge. The second is a *locality* dimension that can be used to indicate the degree of connection between local types of knowledge and global types of knowledge. The greater the degree of connection between local and global knowledge, the greater the power of the curriculum for students in the world of the 21st Century.

Balance and Connection: A Worldly Perspective

We return now to the two questions asked in Chapter 1 and restated at the beginning of this chapter: Do students need strong disciplinary knowledge to cope in today's changing world, or do they need cross-disciplinary, integrated knowledge? Should students learn about local issues, or focus on issues of global significance? In Chapter 1, we suggested that the answer to these questions is "all of these things." We suggested that the curriculum challenge is to get the right balance between disciplinary and integrated knowledge, and the right amount of connection between local and global issues, and we asked how we might do this.

Based on the ten case stories presented, and our synthesis of the findings in the context of other research literature, we argue that the answer to these questions resides in a more holistic view of curriculum than typically exists in schools today. We propose that a Worldly Perspective on curriculum, in which both disciplinary and integrated approaches to solving science-related problems co-exist in a balanced way, provides a powerful model for STEM curricula because it enables science learning to go beyond cognitive, conceptual outcomes by including the social processes and real-world contexts that enable students to become effective citizens. Further, we propose that such a curriculum will demonstrate connection between local issues and global concerns. In other words, we are suggesting that STEM curricula provide a mix of disciplinary and integrated knowledge, set in carefully chosen local problems that can be applied to more global issues. The nature of that mix, finding the point of balance and the degree of connection, is dependent on the particular educational context, and will vary from school to school and from place to place.

Our belief is that schools should seek to provide students with the knowledge that prepares them to be responsible adults and sensible citizens in a rapidly changing global environment, and that STEM curricula have a central role to play. In a recent book (Rennie, Venville, & Wallace, 2012) we explored the contribution that can be made by integrated curricula. We concluded that varying approaches to curriculum integration abound, and their nature and outcomes are dependent on many factors. We endeavored to understand and capture this variety using a Worldly Perspective that, to us, is the crux to understanding the two significant dimensions of curriculum—the balance between disciplinary knowledge and integrated knowledge, and the connection between local types of knowledge and global types of knowledge. We recognize that the point of balance and degree of connection are dependent on educational and curriculum context. What we argue is that the best kinds of STEM curricula will achieve both balance and connection, and so provide students with powerful knowledge to negotiate and improve the community in which they live.

Focus Questions

1. The case stories in this volume demonstrate a range of school contexts in which a teacher, or group of teachers, were involved in an integrated program. Think about the situation in your own school, or a school you have visited recently. What do you expect would be the enabling and hindering factors that would affect the success of an effort to integrate the STEM subjects in this school?
2. What are the advantages and disadvantages that only one teacher working to design and implement an integrated curriculum might experience, compared with a group of two or more teachers?
3. In Chapters 1 and 12, the authors pointed out that there was a variety of curricula described as integrated and that it was possible to put them on a continuum of some kind. The authors disagreed with this kind of thinking, because of the implication of a "better" or "worse" comparison. Now you have considered several case stories, what is your view about the notion of an "integration continuum"?

Suggestions for Further Reading

Rennie, L., Venville, G., & Wallace, J. (2012). *Knowledge that counts in a global community: Exploring the contribution of integrated curriculum*. London: Routledge.

This book expands greatly on many of the issues discussed in Chapters 1 and 12, and discusses in more detail the idea of the Worldly Perspective and the notion of balance and connection in curriculum.

Drake, S. (2000). *Integrated curriculum: A chapter of the ASCD Curriculum Handbook* (pp. 54–63). Alexandria, VA: Association for Curriculum and Supervision.

This handy reference book published by the Association for Curriculum and Supervision provides a general overview of curriculum integration, looks at major trends and issues, includes a question and answer section, and points to various curriculum and other ASCD resources.

References

Aikenhead, G. (2006). *Science education for everyday life: Evidence-based practice*. New York: Teachers College Press.
Au, W. (2007). 'High-stakes testing and curricular control: A qualitative metasynthesis'. *Educational Researcher*, 36(5), 258–267.
Black, P. (1998). *Testing: Friend or foe? Theory and practice of assessment and testing*. London: Falmer Press.
de Brabander, C. J. (2000). 'Knowledge definition, subject, and educational track level: Perceptions of secondary school teachers'. *American Educational Research Journal*, 37(4), 1027–1058.

Gilbert, J. K., Bulte, A. M. W., & Pilot, A. (2011). 'Concept development and transfer in context-based science education'. *International Journal of Science Education*, 33(6), 817–837.

Hall, C. & Kidman, J. (2004). 'Teaching and learning: Mapping the contextual influences'. *International Education Journal*, 5(3), 331–343.

Hurley, M. M. (2001). 'Reviewing integrated science and mathematics: The search for evidence and definitions from new perspectives'. *School Science and Mathematics*, 101(5), 259–268.

Jenkins, E. W. (1999). 'School science, citizenship and the public understanding of science'. *International Journal of Science Education*, 21(7), 703–710.

Kaplan, L. S. (1997). 'Parents' rights: Are middle schools at risk?' *Schools in the Middle*, 7(1), 35–38, 48.

Kelly, G. J., Luke, A., & Green, J. (Eds.). (2008). 'What counts as knowledge in educational settings: Disciplinary knowledge, assessment, and curriculum'. *Review of Research in Education: What counts as knowledge in educational settings: Disciplinary knowledge, assessment, and curriculum* (Vol. 32, pp. vii–x). Thousand Oaks, CA: Sage.

Kruse, R. A. & Roehrig, G. H. (2005). 'A comparison study: Assessing teachers' conceptions with the Chemistry Concepts Inventory'. *Journal of Chemical Education*, 82(8), 1246–1250.

National Research Council. (2011). *A framework for K-12 science education; Practices, crosscutting concepts, and key ideas*. Washington, DC: National Academy Press.

Osborne, J. & Dillon, J. (2008). *Science education in Europe: Critical reflections*. London: King's College, London.

Pang, J. S. & Good, R. (2000). 'A review of the integration of science and mathematics: Implications for further research'. *School Science and Mathematics*, 100(2), 73–82.

Rennie, L. J., Venville, G., & Wallace, J. (2011). 'Learning science in an integrated classroom: Finding balance through theoretical triangulation'. *Journal of Curriculum Studies*, 43(2), 139–162.

Rennie, L., Venville, G., & Wallace, J. (2012). *Knowledge that counts in a global community: Exploring the contribution of integrated curriculum*. London: Routledge.

Rivet, A. E. & Krajcik, J. S. (2008). 'Contextualizing instruction: Leveraging students' prior knowledge and experiences to foster understanding of middle school science'. *Journal of Research in Science Teaching*, 45(1), 79–100.

Ross, J. A. & Hogaboam-Gray, A. (1998). 'Integrating mathematics, science, and technology: Effects on students'. *International Journal of Science Education*, 20(9), 1119–1135.

Stocklmayer, S. M., Rennie, L. J., & Gilbert, J. K. (2010). 'The roles of the formal and informal sectors in the provision of effective science education'. *Studies in Science Education*, 46, 1–44.

Tylack, D. & Tobin, W. (1994). 'The grammar of schooling: Why has it been so hard to change?' *American Educational Research Journal*, 31(3), 453–480.

Venville, G., Wallace, J., Rennie, L., & Malone, J. (2002). 'Curriculum integration: Eroding the high ground of science as a school subject?' *Studies in Science Education*, 37, 43–84.

Wallace, J., Rennie, L., Malone, J., & Venville, G. (2001). 'What we know and what we need to know about curriculum integration in science, mathematics and technology'. *Curriculum Perspectives*, 21(1), 9–15.

Wallace, J., Sheffield, R., Rennie, L., & Venville, G. (2007). 'Looking back, looking forward: Re-searching the conditions for integration in the middle years of schooling'. *Australian Educational Researcher*, 34(2), 29–49.

Young, M. (2008). 'From constructivism to realism in the sociology of the curriculum'. In G. J. Kelly, A. Luke, & J. Green (Vol. Eds.), *Review of Research in Education: What counts as knowledge in educational settings: Disciplinary knowledge, assessment, and curriculum* (Vol. 32, pp. 1–28). Thousand Oaks, CA: Sage.

CONTRIBUTORS

Fiona Budgen is a Lecturer in Education and Primary Mathematics at Edith Cowan University, Western Australia. Dr Budgen's current research interests include teaching and leadership in remote schools, curriculum integration, and the role of language in understanding mathematics.

Rosemary Sian Evans is the Program Coordinator of Science at Balga Senior High School in Perth, Western Australia. She completed her PhD in 2008 at the Science and Mathematics Education Centre at Curtin University, Western Australia. Her primary research concerned scientific literacy and teacher professional learning. Her current position involves developing programs for, and teaching science to, students who are learning English as an additional language.

Susan Joan Gribble is an adjunct Professor of Curtin University, Western Australia. She has been involved with education for over 40 years associated with remote, rural, and urban schools. For the past 25 years, Dr Gribble has been involved in the university sector, teaching undergraduates (Early Childhood Education; Language Learning K–7; Students with Special Needs) and supervising postgraduate students in science, mathematics, and technology education and equity matters. Dr Gribble has experience in international education and internationalization of the curriculum.

Rekha B. Koul is a Research Fellow in the Science and Mathematics Education Centre at Curtin University, Western Australia. Dr Koul's major interests include classroom and school learning environments, gender and cultural issues, and teacher professional development. Currently she is engaged in researching the teaching of engineering and environmental education.

Sheryl MacMath is an instructor and faculty mentor with the Teacher Education Program at the University of the Fraser Valley, Canada. Dr MacMath specializes in lesson planning, assessment and evaluation, mathematics methods, and social studies methods. Her focus is working with both practicing and student teachers to maximize their performance in the field.

Léonie Rennie is Professor of Science and Technology Education in the Science and Mathematics Education Centre at Curtin University, Western Australia. Her current research interests concern science curriculum, with a focus on learning science and technology in integrated and out-of-school contexts, and the promotion of scientific literacy. She is co-author of *Knowledge That Counts in a Global Community: Exploring the Contribution of Integrated Curriculum* (Routledge, 2012).

Rachel Sheffield is a Lecturer in Education and Primary Science at Curtin University. Dr Sheffield's current areas of research interest include investigating teacher change, scientific literacy, and promoting quality discourse in science classrooms.

Grady Venville is Winthrop Professor of Science Education in the Graduate School of Education at the University of Western Australia, Australia. She teaches curriculum and methods to secondary pre-service teachers. Her current research interests focus on curriculum integration, conceptual change, and cognitive acceleration. She is well known as a co-editor of the textbook, *The Art of Teaching Science* (Allen & Unwin, 2004). Her most recent co-authored book is *Knowledge That Counts in a Global Community: Exploring the Contribution of Integrated Curriculum* (Routledge, 2012).

John Wallace is a Professor at the Ontario Institute for Studies in Education, University of Toronto. His teaching and research interests include science teaching, teacher learning, teacher knowledge, curriculum integration, and qualitative inquiry. He is Editor-in-Chief of the *Canadian Journal of Science, Mathematics and Technology Education*. His most recent co-edited/authored books are *Contemporary Qualitative Research: Exemplars for Science and Mathematics Educators* (Springer, 2007) and *Knowledge That Counts in a Global Community: Exploring the Contribution of Integrated Curriculum* (Routledge, 2012).

INDEX

academic tracking 55
access for the disabled project 7, 76–86, **128**; implementation of 77, **78–9**, 80; outcomes of 80–5; research approach 77, 80
Adler, M. 63
Aikenhead, G. 2, 10, 137, 141
American Association for the Advancement of Science 76, 87
Applebee, A.N. 63
approaches to integration vii–viii, 6–8, 24; community-focused or community-based 6, 8, 112–13, 121, **128–9**; cross-curricular 6–7, **126**; project-based 6–7, 34, 124, **125–8**; school-specialized 6, 8; synchronized 6, **127**; thematic, 6, **128**
art, integration with other subjects 7, 25, **26–7**, 50, 106–7, 109, 115
assessment 91, **128**, 138; alignment of teaching and, 73–4, 91–2; authentic 136, 138; common tests 132; concept-mapping 18; formative **78**, **125**, 136; high-stakes 135–6; holistic 117; individualized 121, **129**; learning journal 19–21, 51, 95; portfolio 18, 27; presentations 60–1, 118–9; reflection sheets 16, 18, 20; role-play 117; standards-based 112, 117, 120–1, 135; summative **125**, 136; tests and quizzes 60–1; work books 17; *see also* integrated curriculum, assessment in
Au, W. 135, 141

balance *see* curriculum, balance in
Beale, C. 10
Beane, J. A. 3, 10, 112, 122
Ben-Yehuda, M. 19, 23
Bernstein, L. 53
big ideas *see* curriculum, big ideas in
Black, P. 136, 141
Brabander, C. J. de 137, 141
Brickhouse, N. W. 122
bridge-building project 34–43, **126**; aesthetics 38, 39–40, 42; bridge design 37–8, 42; budget 38–9, 42; cost benefit analysis 42; implementation of 36; outcomes of 42–3
Brown, D. E. 35, 43
Bulte, A. M. W. 137, 142
Burrill, J. 86

case stories viii, 3; Beachville High School 6, 54–62, 124, **127**, 130–2; Chelsea Elementary School 8, 100–10, 124, **129**, 130–4, 138; Gosport Community School 7, 76–86, 124, **128**, 130–4; Greenwich Public School 24–32, 124, **125**, 130, 133, 136; Kentish Middle School 8, 112–121, 124, **129**, 130–1, 133–4, 136, 138; Leaside High School 10–22, 124, **125**, 130, 133, 136; Mossburn School 88–98, **128**, 130–3; Rinkview Public School 64–74, 124, **127**, 130, 133; Seaview Community School 7, 45–53, 124, **126**, 130, 133;

Southern High School 7, 34–43, 124, **126**, 130, 133
Caskey, M. 21, 23
change: adapting to 1; global 1
citizenship vii, ix, 4, 140
civic responsibility 106, 112–3, 120, 138
Clark, E. T. Jr. 4, 9–10
Clarke, S. 88, 98–9
Clement, J. 35, 43
Clement, L. 122
Cobbs, G. 97, 99
collaboration *see* student collaboration; *see* teacher collaboration
community: global 1; links 8–9, 46–8, 96–100, **129**, 133–5; support **126**
connected world vii
connectedness 1; between subjects 22, 76, 85, 97; with community vii, ix, 77, 133
continuum of integration 5
cross-curricular: approach to integration 6–7, **126**; goals 73
Crosthwaite, J. 122
curriculum: balance in ix, 3, 98, 123, 139–40; big ideas in 72, 74, 137; dimensions of 6, 139–40; disciplines in 2, 121, 135, 137–8; expectations 65, 72; integration of vii, viii, 135; *see also* integrated curriculum; learner-centered 4–5, 132; objectives **56**, 61, 65, **114**; organization of 4; purpose of 2, 76; relevance 3, 34–5, 137; structure 2, 34–5; subject-centered 4; teacher-centered 4; tension 112, 120–1; Worldly Perspective on, *see* Worldly Perspective

Dawson, V. 122
decision-making 139
design technology 7; integration with other subjects 12, 36, 50, 64, 109, 115
design: criteria 36; process 13–15
Dillon, J. 137, 141
dimension: knowledge 139; locality 139; multiple 1
Doerr, H. 21, 23, 33
Donohue, M. H. 109, 111
Drake, S. 5, 10, 74, 141
Driver, R. 35, 43
Durán, R. P. 122

Earl, L. 5, 10
education: for citizenship *see* citizenship; place-based 2

electricity and energy use project 6, 54–62, **127**; implementation of 55–9; objectives 161; research approach 54; success in 60–1
engineering 34–43, integration with other subjects 34, 36; learning 36–40; mockups 14; prototypes 13; social aspects of 42
English *see* language arts
environmental education *see* snakes project; *see* water quality project; environmental responsibility 7, 100, 106, 108, 120, 138
explicitness in teaching 74

Flihan, S. 63
Flowers, N. 54, 63
focus: arguments for, 72; confusion 73; diffuseness of, 73; uncertainty 73
Fogarty, R. 5, 10
Fogelberg, E. 53
funding 100, 105, 107–8, **129**

Gardner, H. 2, 10
geography integrated with science 55, **56–9**, **128**
Gilbert, J. K. 137, 142
Gilbert, J. K. 137, 142
global change 1; community 1; context 43; issues/problems 139–40; world 3
Good, R. 135, 141
Grable, A. 10
grammar of schooling 134
Green, J. 112, 122, 135, 142
Gruenewald, D. A. 3, 10
Gunstone, R. F. 98

Hall, C. 134–5, 142
Hall, G. E. 88, 98
Hargreaves, A. 5, 10
Harkness, S. S. 22
Hatch, T. 2, 10
Hernández-Gantes, V. M. 86
Hiller, B. 53
Hodson, D. 2, 10
Hogaboam-Gray, A. 32–3, 72, 75, 136, 142
holistic perspective ix, 123, 136, 139–40
Hord, S. M. 88, 98
horticulture 50–1, **126**, 133
Hurley, M. M. 136, 142

ice hockey project 24–32, **125**; building a rink 29–31; outcomes of 31–2; program for **26–7**; research approach 24; team schedule 27–9

Indigenous students 45–8, 52, 109

integrated curriculum: approaches to *see* approaches to integration; arguments for 3; assessment in 17–22, 25, 31, 35, 106, 117–21, 136; attributes of 2, 3, 124; barriers to 76, 85; challenges 54, 72, 88; context 4, 31–2; definition of 3–6; enablers for **124–129**, 130–135; focus in 72–4; implementation of 4, 12, 25, 54, 124; meaning of viii; measurement in 138; outcomes of vii, 15–22, 31–2, 42–3, 51–2, 60–1, 72–3, 97–8, 104–10, 115–17, 119–21, 136; popularity of vii

interdisciplinary 137–8; curriculum 2, 4–5, 7–8, 109, 138; goals 64; objectives, 136; teams vii, 124

Jacobs, H. 5, 10
Jenkins, E. W. 137, 142
Johnson, G. M. 55, 60, 63
Johnson, J. 43
Johnson, M. D. 34, 44

Kaplan, L. S. 135, 142
Kelly, G. J. 112, 122, 135, 142
Kenney, J. L. 109, 111
Kidman, J. 134–5, 142
knowledge: application 20–2, 29, 40–2, 73; balance between disciplinary and integrated 123, 138–40; connections between local and global 3, 4, 139; cross-disciplinary 3, 138; disciplinary 3, 137–8; integrated 3, 137–8; powerful 137–9; professions-based 34; status of 137, 139; transfer 20–2, 29, 40–2, **57**, 58–61, 72, **125**
Krajcik, J. S. 2, 10, 137, 142
Kruse, R. A. 136, 142

language and horticulture case story 46–53; horticultural program 50–1; implementation 47–9; integrated program 49–50; outcomes of 51–2; research approach 46; school community 45–7

language arts: and action 21–2; integration with other subjects **26–7**, 49–52, 65, 88, **91**, 106, 115, **126**, 130

Lavay, I. 19, 23
Leach, J. 76, 87
leadership ix, 88, **129**, 130–32; administrative **126**; shared 88, 94, 97–8, **128**, 130–1; supportive 47, 124; tensions 96–7; transformative 130

learning: assessment of *see* assessment; conceptual **129**, 138; context, importance of 29–31; cooperative 17, **127**; community 98, 113–5, 118, 121, 124, 130; environment 108, 131; in integrated curriculum 16, 31–2, 138; meaning-making as 4; and motivation 19, 21; reflection on 19; *see also* students

Lederman, N. G. 35, 44
Levinson, R. 86
Lewis, J. 76, 87
Linchevski L. 19, 23
literacy: language 7–8, 45–53, 123, 130; scientific 109, 113, 136, 138–9; technological 136, 138–9
Lloyd, D. 2, 10
local issues ix, 3, 124, 139–40
Lock, R. 122
Luke, A. 112, 122, 135, 142

Malone, J. 4, 11, 134, 139, 141–2
marketing *see* toy project
mathematics concepts: graphing 67–8; proportions 27–9, 40; ratio and scale 29–31
mathematics, integration with other subjects 12, 25, **26–7**, 36, 50–1, 64, 88, **91**, 106, 114
McKee, J. A. 53
Meier, S. 97, 99
mentoring 83, 134
Mertens, S. B. 54, 63
metawareness in teaching and learning 73
Meyer, H. 22
middle-school 3, 77, 112
Militana, H. P. 109, 111
Miller, J. 2, 10
misconceptions **125**; in mathematics 28–9; in science 35, 42
model house project 88–98, **128**, 130; house construction 92–4; implementation of 89–92; outcomes of 97–8; project leadership 89; research approach 92
motivation: connection with skills 16; *see also* students, motivation

Mulhall, P. F. 54, 63
multidisciplinary: curriculum 4–7, 135, 138; problems 2

National Research Council 76, 87, 137, 142
Nazir, J. 3, 10
Nicol, M. 97, 99
Niess, M. L. 35, 44
Nuthall, G. 63

O'Donoghue, T. 88, 98–9
O'Loughlin, M. 3, 10
Ogle, D. 53
Osborne, J. 137, 142
outcomes, clarification of **125**; *see also* integrated curriculum, outcomes of

Pang, J. S. 135, 141
parental: expectations 135, 137–8; support 46–7, 52, 55, 96, 103
Pedretti, E. 3, 10
Pendergast, D. 23
Perkins, D. N. 20, 23
perspective, Worldly *see* Worldly Perspective
Petrosino, A. J. 120, 122, 124
Pilot, A. 137, 142
planning process *see* design process
Powers, A. L. 110
problem solving vii, 16, 19, 28, 34, 39, 72, 138
problems: community 8; discipline-based 2; global 3; real-world vii, viii, 5, 35–6, 42, 104–121, 139; STEM-related viii, ix

real world problems *see* problems, real world
reflection *see* learning, reflection on; *see* teaching, reflection on
reinforcement 55–62
Rennie, L. 4, 6, 10–11, 43, 74, 97, 99, 110, 120, 122, 124, 130, 132, 134–7, 139–142
research program 3–5, 8–9
Rivet, A. E. 2, 10, 137, 142
Robertson, A. 10
robots 80, 82–3, 85
rockets and submarines project 12–22; bottle rocket 12–3, 15, 17–8; outcomes of 15–22; pull-along toy 14, 16, 18, 20; research approach 12; rocket cars 13, 15, 18; submarine 14, 19–21
Roehrig, G. H. 136, 142

Rogers, B. 2, 10, 34, 44
role play 36, **102**, 103, 114–15, 117, 119
Ross, J. A. 32–3, 72, 75, 136, 142
Rushworth, P. 35, 43
Russell, T. 98
Ryan, J. 5, 10

Satz, P. 53
scaffolding 35
Schoenfeld, A. H. 19, 23
school: administrative support 76, 80–3, **128**, 132; case story, *see* case story schools; context 134–5; culture of 134–5, 138; education 1; timetable, *see* timetable
school-community project *see* snakes project; *see* water quality project
science concepts: ecology 100, 102, 104–5, 113, 115; electricity 56–61; energy 56–61; food webs 103, 115–6; forces 35–6, 41–2; levers 65–6; salinity 115–6, 120; water cycle 121
science, integration with other subjects 12, **26–7**, 36, 50, 55, **56–9**, 64, **78–9**, 88, **91**, 115, 120–1, **127**
Scott, D. 2, 11
Sfard, A. 19, 23
Sheffield, R. 43, 97, 99, 120, 122, 124, 135, 142
Skalinder, C. 53
Smith, G. A. 3, 10
snakes project 100–10; community night 101–3, 109, **129**; community survey 102–3; funding 100, 108–9; implementation 101–3; outcomes 104–110; research approach 101; wildlife center **102**, 105–6, 108
social responsibility 105–6, 138
social studies, integration with other subjects **26–7**, 64–5, **78–9**, 88, **91**, 119–20, 130–1
Sowder, J. 122
Squires, A. 35, 43
Stallworth, J. 22
STEM: career choice 137; curricula 4, 123, 135–6, 139, 140; subjects 1, 9, 137, 139
Stinson, K. 22
Stocklmayer, S. 137, 142
STSE 3
students: "at risk" 55; attitudes 17, 19; collaboration between 17, disaffected 81–2, 84–5; engagement 15, 19, 71–2,

93, 95–6, 103; group or team work 13–15, 19, 36, 60, 67, 70–2, 93, 118, 136; independence 15–17, 22, 51, 119, **129**; interest 15, 84, 103–4, **125**; motivation 15–17, 19, 25, 51, 54, 104–7, 110, **125**, **127**, 136; responsibility for learning 16; self-efficacy 60–62
subjects: boundaries between 7, 135

teacher: collaboration 55–6, 65, **79**, 80, 85, 88, 91, 95, 108, 124, **127**, 130, 132; effort 4, 6; inexperience 83, **128**; isolation 83–4, 124, **128**, 132; knowledge **126–7**, **129**, 130–1, 134–5; lack of time 94–6, 107, 132; modeling 13; reflection 31, 61–2, 69–70, 85, 108
teaching styles 105
teaching: commitment to **128**, 130; lack of time 94–6, 107; out of field **129**, 136; planning time 55, 61, 90–2, 105–7, 109–10, **125**, **127–8**; space **125–6**; teams 77, 88, 101, 124, **125**, **127**, 131–2; workload 96, 107, **125**, **127**, 132
technology *see* design technology
Thomson, N. 109, 111
timetable 2, 80–1, 115, **125–6**, **128–9**, 132–4
Tobin, W. 134, 142

toy project 64–74, **127**; implementation and objectives of 64–65; construction 70–1; experiments with levers 65–6; outcomes 66, 72–4; research approach 64; schematic designs 68–9; surveys 66–7; toy fair 71–2
transdisciplinary curriculum 4–5, 8, 137
Tylack, D. 134, 142
Tyler, R. W. 2, 3, 11

unidisciplinary 4

values 7, 106–7, 112–21
Venville, G. 4, 6, 10–11, 43, 74, 97, 99, 120, 122, 124, 130, 132, 134–6, 139–42
Vygotsky, L. 19, 23

Wallace, J. 2, 4, 6, 10–11, 43, 74, 97, 99, 120, 122, 124, 130, 132, 134–6, 139–42
water quality project 112–121, **129**; approach to assessment 117–121; funding 113; implementation 113–5; outcomes 115–17, 121; research approach 113
White, G. 98
Wood-Robinson, V. 35, 43
Worldly Perspective ix, 123, 139–40

Young, M. 137, 142